Math Made Simple

Grade 4

Written by Kaye Furlong
Illustrated by Don Ellens

Notice! Copies of student pages may be reproduced by the teacher for home or classroom use only, not for commercial resale. No part of this publication may be reproduced for storage in a retrieval system, or transmitted in any form or by any means–electronic, mechanical, recording, etc.–without the prior written permission of the publisher. Reproduction of these materials for an entire school or school system is strictly prohibited.

FS-23204 Math Made Simple Grade 4
All rights reserved–Printed in the U.S.A.
Copyright © 1997 Frank Schaffer Publications, Inc.
23740 Hawthorne Blvd.
Torrance, CA 90505

Introduction

Math plays a vital role in everyone's life on a daily basis—in a variety of situations. It is, therefore, absolutely a necessity that children develop, understand, and learn to apply math skills.

What is symmetry?

How will math make me better understand sports? cooking? shopping?

How are decimals and fractions related?

How is perimeter different from area?

These and other questions are children's attempts to make sense of mathematics concepts.

Math Made Simple has been designed to help students develop a basic understanding of math concepts and to help them practice skills and algorithms related to those concepts. The activities presented in this book help students learn to apply these skills and concepts in a variety of problem solving situations.

The objective of *Math Made Simple* is to help all students succeed in math. In order to ensure success for all learners, the activities in this book are presented in a variety of formats.

This book is divided into five sections. In the front of each section are Teacher Resource pages. These pages feature many exciting extension activities that you can guide the students to do. Another fascinating aspect of the Teacher Resource pages is the School-Home Connection activities. These activities provide a great way for students to apply the skills and concepts presented to situations at home. These kinds of activities are a great way to show students and their families that opportunities for learning math can arise at home as well as at school and that learning math can be interesting, fun, and valuable.

Behind the Teacher Resource pages are a lot of fun and stimulating activity pages students can complete to learn the important math skills and concepts featured in the section. The activities include playing games, decoding messages, solving riddles, working puzzles, coloring, and much more. These activities provide yet another exciting means for students to better understand the concepts and skills presented in each section.

The concepts covered in *Math Made Simple* are basic to most fourth grade mathematics programs. Students will develop conceptual understanding of and will practice skills relating to the following mathematical concepts: numeration; addition and subtraction of whole numbers, decimals, and fractions; multiplication and division of whole numbers; geometry; measurement; decimals and fractions; and probability and graphing.

To reinforce the math topics presented in this book and to help students gain a greater understanding of these topics, prepare (or have the students prepare) math journals. These provide students with places to write down the thinking processes or steps that they have used to solve problems. They are also wonderful places for students to record any interesting mathematical discoveries they make. Let the students share their journals with each other and with you when applicable.

Another valuable tool you and the students can make are erasable cards (i.e., laminated pieces of paper). These are great for students to use when doing group activities. Students can write down answers to problems using crayons and hold them up. This enables you to quickly check to see who understands the concept being presented. Smaller groups can then be created to reteach a particular skill.

Regardless of your reasons for implementing *Math Made Simple*, you will be delighted as you watch your students discover how interesting and fun learning math can be!

Whole Numbers

The activities in this section (pages 1–33) are perfect to help students understand the number system, especially place value and the relationship of numbers. This understanding will help provide students with a foundation they can use to develop skills in computation. A solid knowledge of place value also provides a good foundation to help students order and compare numbers. Be sure to provide students with ample opportunities to work with manipulatives and complete several examples with your guidance. Also be sure to point out to students how they can use their new skills in everyday life. Daily experiences help students see the connection between math and real life.

CONCEPTS

The ideas and activities presented in this section will help students explore the following concepts:
- place value to millions
- comparing and rounding numbers
- Roman numerals
- adding and subtracting 2- to 5-digit numbers
- estimating sums and differences
- problem solving
- multiplying 1- to 2-digit numbers
- dividing with and without remainders

Numeration

DIFFERENT WAYS TO SHOW NUMBERS — Group Activity

On separate cards, write numbers in digits (4,987), in words (four thousand, nine hundred eighty-seven), and as expanded numerals (4,000 + 900 + 80 + 7). Have the cards ready for small group lessons in which students match the cards. You could also use the cards for a journal experience. For this, have the students pick three ways to write numbers and show those three ways in their journals.

MORE THAN, LESS THAN, OR EQUAL TO — Group Activity

Provide each student with 3 small squares of construction paper (3 different colors). Have the students write signs for greater than (>), less than (<), and equal to (=) on the cards using the colors you determine. Orally (or on the board), name different numbers such as 2,804 and 2,408. Have the students indicate more than, less than, or equal to by holding up the correct cards.

ROUNDING — Group Activity

Have the students use their erasable cards to show the rounded version of a number you state. For example, if you write 471 on the board, students show 470 if asked to round to the nearest 10, or 500 if asked to round to the nearest 100. Make a chart and post it in your classroom stating that numbers rounded to the nearest 10 end in one zero, and those rounded to the nearest 100 end in two zeros. Later, you can add the fact that numbers rounded to thousands end in three zeros to the chart. This will save a lot of confusion until students are really in command of this concept.

FS-23204 Math Made Simple • © Frank Schaffer Publications, Inc.

PLACE VALUE BOX

Game

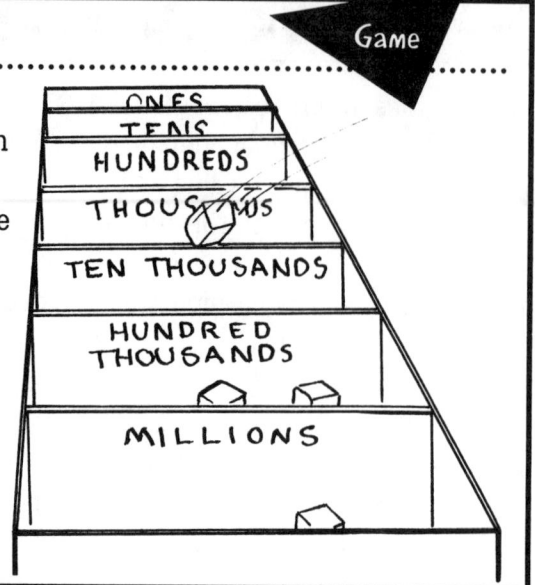

Tape 7 shoeboxes together. On each box, write a place value from ones to millions. Have the students take turns standing on a taped line and gently tossing 10 cubes into the boxes. Each student reads the number he or she has made by taking out the cubes that landed in the boxes. For example, if 3 cubes landed in the thousands box, 2 in the hundreds, none in the tens, and 4 in the ones, the number would be read 3,204. Students can compete for the largest number in their group. Or, have students keep a running total of their numbers to compete for the largest number. You and the students could also make a graph of the totals achieved by all the different groups. Students can use the graph to write story problems.

ORDINAL NUMBERS

Class Activity

Have students place cards with ordinal numbers in order on the chalk tray, or they can hold the cards and stand in order. Students can also line up and call off their order numbers from first through the last ordinal number in the class. While in order, you could also have students state their places and ask questions. For example, you start by asking who is in sixth place. The sixth student answers and asks, "I am sixth. Who is four places behind me?" This student then answers and asks, "I am tenth. Who is seven places behind me?", etc.

Homework

School-Home Connection

Have each student find 10 large numbers in a newspaper. Students should cut the numbers out and paste them on a piece of paper. Beside each number, students should write an explanation of how the number was used.

When the papers are returned to school, you and the students can create story problems to go with the numbers.

ROMAN NUMERALS

Class Activities

Post a "Roman Numerals" chart showing Roman numerals and their equivalents in standard form (i.e., XII = 12). You could develop the chart with the students as you teach or review Roman numerals. Try the activities below to help students learn Roman numerals.

- Working on the board or overhead, have students write Roman numerals, as many as possible, without using the chart.
- Students should make a copy of the chart for their journals.
- Have the students write various Roman numerals on cards to challenge groups of students to read. You could place the cards in a center and have students write the standard form of each Roman numeral.
- You could have the students write story problems that must be figured out using Roman numerals.

Millions and Millions

Write each number in the place value boxes.

Millions			Thousands			Ones		
	6							

1. 65,703,516

2. 841,654,812

3. 140,372,166

Write the numbers.

4. five million, seven hundred thousand, nine hundred two

   ____ , ____ ____ ____ , ____ ____ ____

5. 20,000,000 + 50,000 + 7,000 + 600 + 80 + 4

   ____ ____ , ____ ____ ____ , ____ ____ ____

6. one million, nine hundred six thousand, two hundred six

   ____ , ____ ____ ____ , ____ ____ ____

7. 2,000,000 + 600,000 + 40,000 + 300 + 20 + 1

   ____ , ____ ____ ____ , ____ ____ ____

8. nine million, two hundred two thousand

   ____ , ____ ____ ____ , ____ ____ ____

9. twenty-five million, eight hundred one

   _____

10. 19,000,000 + 50,000 + 900 + 5

    _____

11. 400,000,000 + 3,000,000 + 90,000 + 70 + 7

    _____

12. seventeen million, sixty-four thousand, ninety-six

    _____

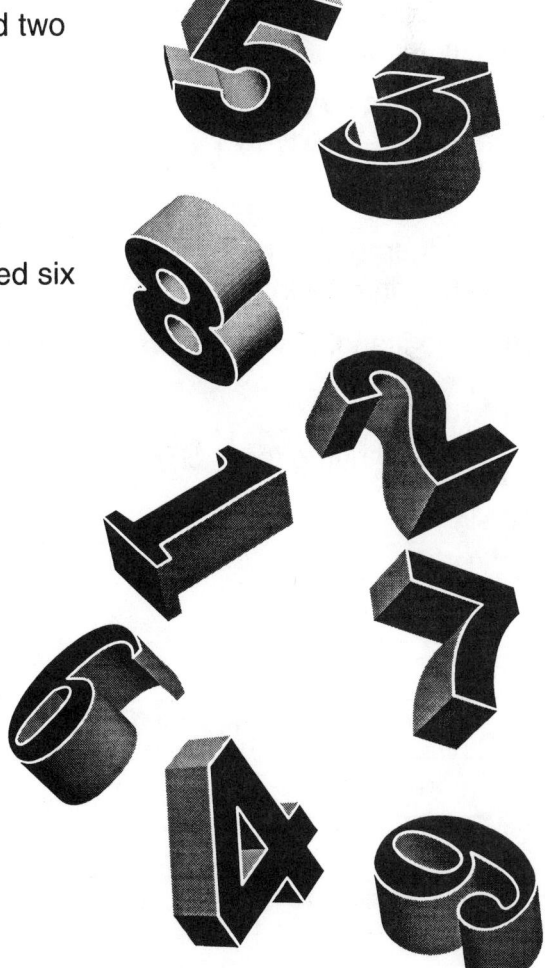

Place value—millions 3

Big Fish

Write the numbers in the puzzle.

Across

1. eight hundred eighty-nine thousand, six hundred twelve
6. forty-three thousand, eleven
8. eight thousand, four hundred forty
9. three hundred million, eight hundred sixty-five thousand, seven hundred seventy-two
10. thirty-nine thousand, thirteen
11. six thousand, five hundred eighty-eight
14. forty million, two thousand, one hundred sixty-two
16. thirty thousand, one hundred nine
18. five hundred fifty thousand, one hundred twenty-three
19. fifty thousand, seven hundred eighty-six

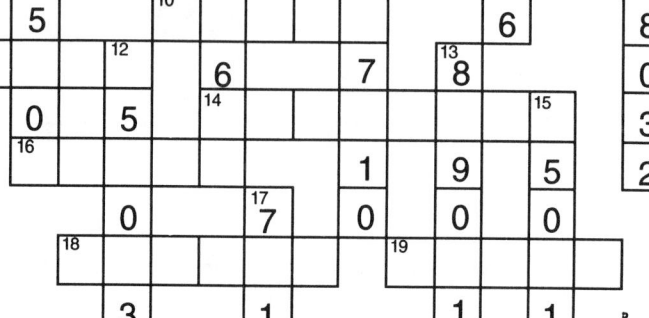

Write the word name for each number.

Down

2. _____
3. _____
4. _____
5. _____
7. _____
12. _____
13. _____
15. _____
17. _____

Place value—millions

Let's Compare

Write > or <.

A.   10,769 ◯ 11,200   177,204 ◯ 116,791   97,213 ◯ 107,213

B.   86,190 ◯ 84,993   981,203 ◯ 918,203   72,218 ◯ 75,003

C.   198,221 ◯ 98,221   264,193 ◯ 266,001   83,103 ◯ 80,793

D.   656,183 ◯ 665,183   199,820 ◯ 198,993   55,186 ◯ 54,939

E. Write the numbers in order in the chart below.

| 756,802 | 694,213 | 820,000 | 971,246 | 378,461 | 192,874 |
| 471,622 | 216,606 | 92,813 | 87,214 | 63,744 | 55,806 |

Big	Bigger	Biggest
1.	5.	9.
2.	6.	10.
3.	7.	11.
4.	8.	12.

F. The smallest number is _____.

   The largest number is _____.

G. Write the number that is 10 less than the smallest number. _____

   Write the number that is 100 more than the smallest number. _____

   Write the number that is 1,000 more than the largest number. _____

Comparing numbers

Round and Round

In each number below, circle the number that tells you if the number should round up or stay the same. Round each number.

A. **10**

85◎6◎ _____

47 _____

293 _____

64 _____

327 _____

B. **100**

467 _____

821 _____

299 _____

1,371 _____

5,614 _____

C. **1,000**

5,764 _____

8,213 _____

9,621 _____

2,473 _____

1,121 _____

D. Round to the nearest thousand. Color the balls when the numbers round up.

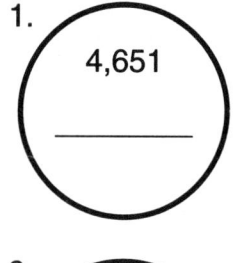
1. 4,651 _____

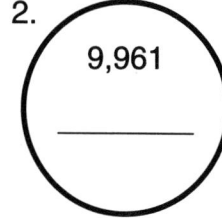
2. 9,961 _____

3. 7,208 _____

4. 2,976 _____

5. 899 _____

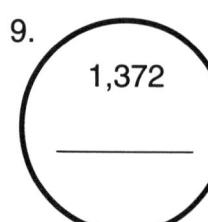
6. 3,479 _____

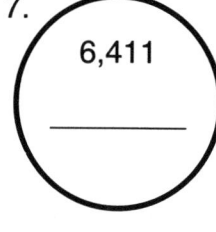
7. 6,411 _____

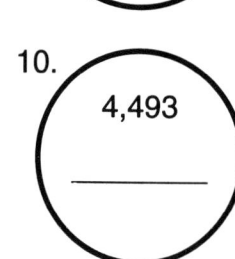
8. 5,743 _____

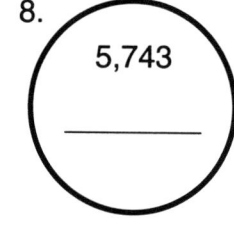
9. 1,372 _____

10. 4,493 _____

11. 7,510 _____

12. 6,821 _____

13. 2,093 _____

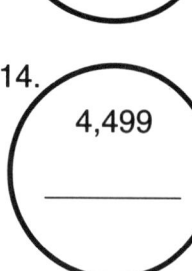
14. 4,499 _____

15. 8,803 _____

E. Answers for tens need to end in _____ zero(s).

Those for the nearest hundred end in _____ zero(s).

Answers to the nearest thousand end in _____ zero(s).

Another Kind of Number

A. Finish the chart.

I = 1	VII = ___	XL = 40	C = 100	DC = 600
II = 2	VIII = ___	L = 50	CC = ___	DCC = 700
III = 3	IX = 9	LX = 60	CCC = ___	DCCL = ___
IV = 4	X = 10	LXX = ___	CCCL = ___	DCCC = ___
V = ___	XX = ___	LXXX = ___	CD = 400	CM = 900
VI = 6	XXX = ___	XC = 90	D = ___	M = 1,000

B. To make the numbers between 10–13, add ones (I) to ten (X).

Write 11, 12, and 13 in Roman numerals. 11 = _____ 12 = _____ 13 = _____

Fifteen is 10 + 5, or _____.

For 16, 17, and 18, add ones to 15. 16 = _____ 17 = _____ 18 = _____

For 14, subtract 1 from 15 by putting the one in front of 5. 14 = _____

For 19, subtract 1 from 20 by putting the one in front of the second 10. 19 = _____

C. Following the rules above, write the numbers.

XXII = _____ XXXIII = _____ LII = _____ CDXCI = _____

CL = _____ CLX = _____ DL = _____ DCLI = _____

D. Draw lines to match.

CCCX	560	XVI	24
XXXIV	8	CCXC	16
DLX	310	DX	19
VIII	34	XIX	290
XIV	805	XXIV	510
DCCCV	14	MMCIX	740
MIX	1,009	DCCXL	2,109

Roman numerals

Adding and Subtracting Whole Numbers

MISSING ADDENDS AND OPERATION SIGNS
Class Activity

On an overhead projector, write number sentences with missing numbers and missing operation signs. For example, 14 ○ □ = 12 – 6. Students find the answer to 12 – 6 (6) and determine what operation and number are needed on the left to make the sentence true (14 – 8). Students can then fill in the missing information. Some problem solving strategies may need to be developed, such as working from the known (12 – 6 = 6) to the unknown. Ask students how they can get an answer of 6 from 14. (They must subtract to get a lower number. They subtract to find what number gives a difference of 6 when subtracted from 14; 14 – 6 = 8, so 8 must be the number needed.) After writing the answers into the number sentence, students should check to see if the answer makes the sentence true.

ADDING TWO-DIGIT NUMBERS
Mental Math

Have students add the ones' column of an addition problem you have written on the board. For example, if your problem is 46 + 27, the students should add the 6 and 7 and think of 13. They now know the answer will end in a 3. They need to remember that they have a one to regroup when adding the 4 tens and 2 tens, or 4 + 2 + 1 = 7. After practicing orally, the students should show their numbers on their erasable cards so you can quickly see who has the idea and who needs more practice.

COMPARING SPORTS NUMBERS
Writing Activity

Many students will enjoy using facts from the sports page to make up story problems. They can also use advertisements to create addition and subtraction problems by subtracting to find the amounts saved when items were bought on sale, etc.

DESCRIBING PROCESSES
Writing Activity

Have the students write their own definitions of addition with and without regrouping in their journals. They can describe how they add, or what they think it means. You will get a variety of responses, but you will know if the students have the general idea. Do this again with subtraction, multiplication, and division. Encourage students to write complete responses showing their thinking. They can do this either by reading the responses of other students or by writing them on the overhead. This way, students can discuss the responses, add to them or change them, or compliment a student's response as very clear and complete.

MENTAL MATH ROUNDING

Class Activity

When teaching rounding, be sure to put up the chart mentioned in the activity on page 1, "Rounding". This chart will help students see that when you round to the nearest ten, the answer must end in one zero; nearest hundreds need two zeros; and nearest thousands need three zeros. You can also display a number line showing 5 or more rounding to the next ten, 50 or more rounding to the next hundred, and 500 or more rounding to the next thousand. A great activity to fill short time periods is to call out numbers for the students to round and add together. Let them use their erasable cards to show their answers. For example, call out 457 + 371. Students should think 500 + 400 and show 900. You can do the same thing with subtraction. (For example, 842 − 567 = 800 − 600 = 200).

COMMUTATIVE PROPERTIES

Class Demonstration

Have students add four different numbers together. Then have them move the numbers around and add them again. Have students repeat this four times. Ask students if it matters in what order the digits are added. (no) Then give students a subtraction problem. Ask students what happens when the smaller number is on top. (They can't subtract.) Write the word *commutative* on the board. Explain to students that addition is commutative and subtraction is not. Having just done the experiment, have students write a definition for *commutative* in their journals. Help them by discussing their answers. Allow students to change their answers if they missed the idea.

TEAM RIDDLES

Game

Divide students into teams. Have each team think up four riddles such as "I am less than 100, but more than 30. The sum of my digits is 7. What number am I?" The other team tries to solve the riddle. It may come up with the answer of 43, which follows the clues. However, the answer that the other team is looking for is really 34. The team, therefore, must continue asking questions until it gets the right answer. Keep track of the number of answers tried by the other team. The team posing the riddle gets the number of guesses as its points in that round. Students will soon begin to form generalizations or ask themselves questions such as "If the sum is 7, and the answer isn't 43, what else can it be? (34, 52, etc.)." Students should also recognize why the answer can't be 25 (below 30). This activity promotes creative thinking and logic.

FS-23204 Math Made Simple ▪ © Frank Schaffer Publications, Inc.

BASEBALL MATH AND RELAYS — Games

Divide the class into two teams. Designate each corner of your room as first base, second base, third base, or home. The first student from a team draws a problem that has been written by you or another student on a card. He or she then writes the problem on the board and solves it. If correct, the student advances to first base. The next player comes up and repeats the process. When a player is wrong, the team receives an "out," and the players in front of him or her don't move ahead. After four correct answers, the team receives a point as the first runner reaches home. Continue until three outs are reached. At this point, it is the other team's "bat." (You could also have both teams participate and move around the bases at the same time, ignoring outs, which encourages more sustained group interest.)

For relays, divide students into 2, 3, or 4 teams. A student from each team can come to the board to solve a given problem. You can give a point to the first team member that is correct, or students can keep their answers covered and show them at the same time, and a point can be awarded to all teams with the correct answer.

LINE UP — Manipulative Activity

Many students have trouble trying to line up numbers in the correct columns when adding or subtracting, especially as numbers get longer. To help students, have them turn their papers sideways so they can write numbers between the lines for columns. Or, you could provide ½-inch graph paper for the students to use.

School-Home Connection — Homework

Tell students to cut out advertisements from the newspaper. Students use these ads to find a certain number of items that come closest to a given amount of money. (Have parents decide on the number of items and the amount of money.) Students can also use the ads for questions their parents ask them such as "How much do you save by buying the item on sale? How much could you save if you bought three of that item on sale?", etc. Have students create problems for class use.

Name_____

Add It Up!

Color the sums.

red = 1,400–7,900   green = 8,000–16,000   yellow = over 16,000   blue = under 1,400

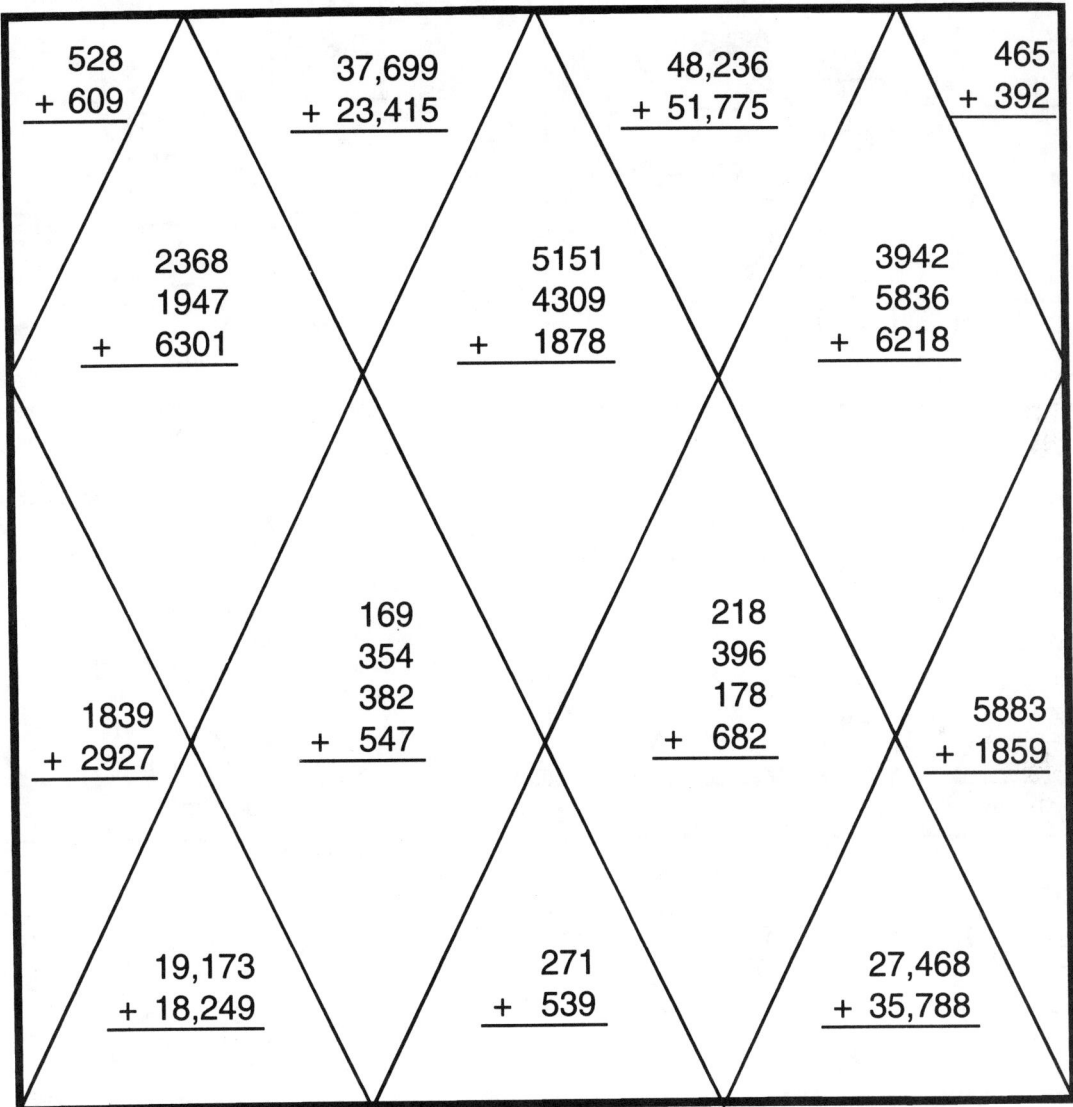

Solve.

1. Julie and her dad drove across the country on vacation. They drove 349 miles, 276 miles, and 362 miles in three days. How many miles did they drive in all?

   _____

2. Our postman delivers mail five days each week. He delivered 249 pieces of mail on Monday. On the other days, he delivered 782 letters and 381 magazines. How many pieces of mail did he deliver in that one week?

   _____

2- to 5-digit addition (2 or more addends)

Shapely Sums

Working diagonally, add the numbers shown by the arrows.
Write >, <, or = to compare the sums.

1.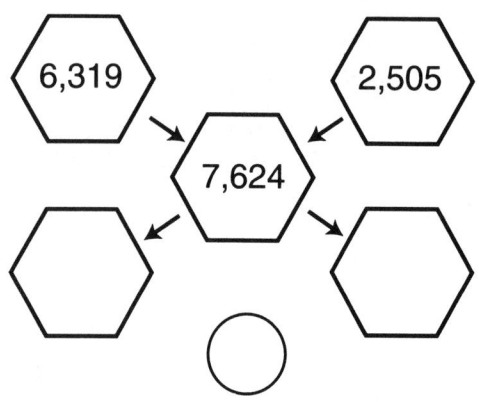

2.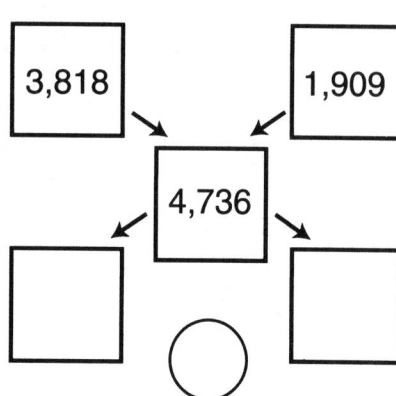

3.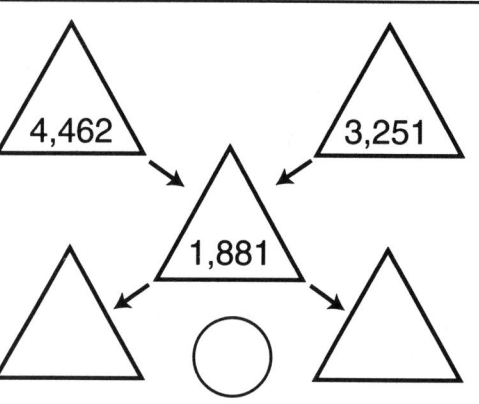

4.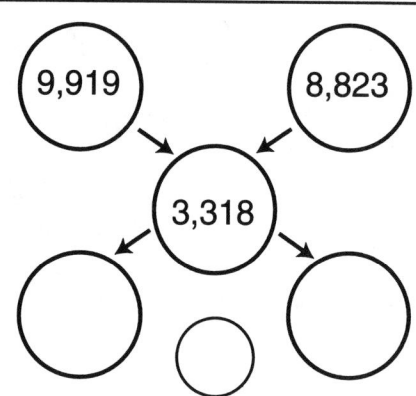

5.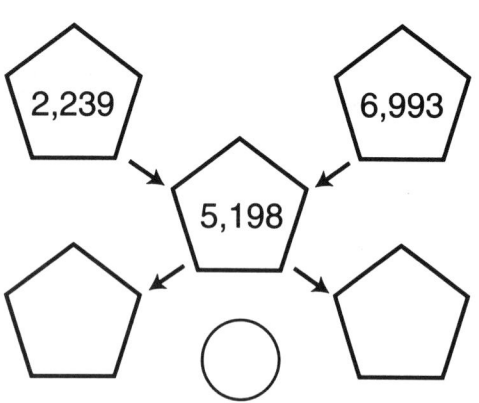

6.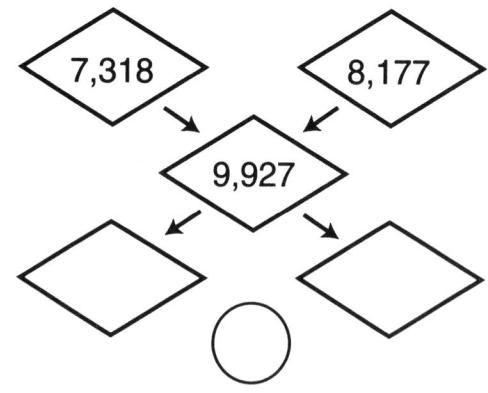

Write the name of the shape in each puzzle.

1. _____   2. _____   3. _____
4. _____   5. _____   6. _____

Sport Shopping

Fill in and total the receipts. Be careful! Sometimes the amount is more than one.

A.

Jeff's Sporting Goods		Date _____	
Amount	Item	Price per item	Total
2	baseballs	_____	_____
1	bat	_____	_____
1	glove	_____	_____
		Grand Total	_____

B.

Jeff's Sporting Goods		Date _____	
Amount	Item	Price per item	Total
2	tennis rackets	_____	_____
3	cans tennis balls	_____	_____
		Grand Total	_____

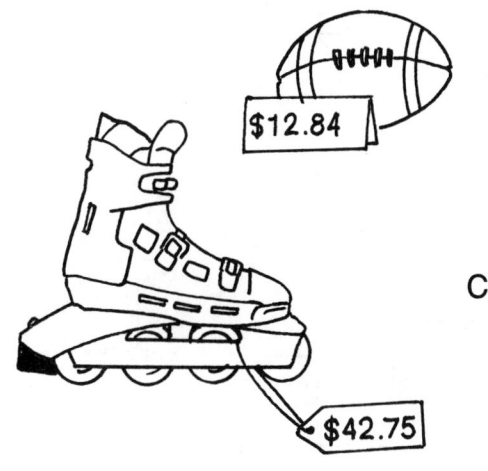

C.

Jeff's Sporting Goods		Date _____	
Amount	Item	Price per item	Total
1 pair	in-line skates	_____	_____
2	cans tennis balls	_____	_____
		Grand Total	_____

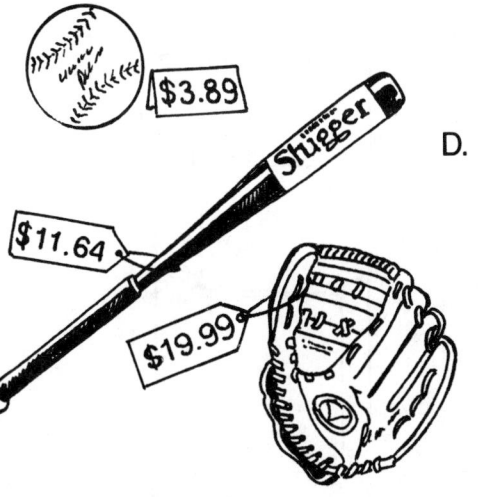

D.

Jeff's Sporting Goods		Date _____	
Amount	Item	Price per item	Total
2	footballs	_____	_____
3	baseballs	_____	_____
4	cans tennis balls	_____	_____
		Grand Total	_____

Addition of money

Zeros Are Important

A. Subtract. Add to check your answers.

1. 207   64
   −143   +143
    64   207

2. 100
   − 64   +_____

3. 602
   −314   +_____

4. 5,007
   −2,106   +_____

5. 4,109
   −1,174   +_____

6. 3,000
   −2,674   +_____

7. 9,008
   −3,962   +_____

8. 12,003
   −10,666 +_____

9. 70,405
   −21,510 +_____

B. Write the problems below in column form and subtract.
   Write the answers in correct order from smallest to largest.

1. 560 − 392 = _____
2. 4,300 − 571 = _____
3. 5,000 − 297 = _____
4. 640 − 399 = _____
5. 7,040 − 2,607 = _____
6. 1,007 − 352 = _____

largest

smallest

1. _____ 2. _____ 3. _____ 4. _____ 5. _____ 6. _____

2- to 5-digit subtraction

Name_____

Puzzle Fun

Subtract.

1. 4,927
   − 1,419

2. 3,076
   − 1,034

3. 7,878
   − 3,921

4. 2,104
   − 1,003

5. 9,875
   − 5,999

6. 8,989
   − 7,979

7. 3,416
   − 1,027

8. 9,999
   − 8,888

9. 45,689
   − 26,553

10. 97,532
    − 65,965

11. 11,621
    − 11,401

12. 6,215
    − 3,147

13. 38,654
    − 3,527

14. 6,868
    − 4,099

15. 84,230
    − 65,432

FS-23204 Math Made Simple ▪ © Frank Schaffer Publications, Inc.

4- to 5-digit subtraction **15**

Name_____

Let's Eat

Write a bill for the food you buy. Find your change.

Mac's Cafe

Mini Burger $2.99	Garden Salad $2.19	Milk $1.02
Happy Burger $3.45	Veggie Burger $3.08	Cola $1.17
Super Cheeseburger ..$4.15	French Fries $1.66	Orange Juice $1.08
Jumbo Burger $5.87	Onion Rings..............$1.75	Lemonade $1.24

A. Mac's Cafe

1 Happy Burger _____
1 Onion Rings _____
1 Lemonade _____
    Total _____
Amount Received $10.00
    Change _____

B. Mac's Cafe

2 Garden Salads _____
2 Milks _____
    Total _____
Amount Received $10.00
    Change _____

C. Mac's Cafe

2 Jumbo Burgers _____
2 French Fries _____
2 Colas _____
    Total _____
Amount Received $20.00
    Change _____

D. Mac's Cafe

3 Veggie Burgers _____
1 Milk _____
2 Orange Juices _____
    Total _____
Amount Received $15.00
    Change _____

On the back of this page, list the food you would buy for a party of four friends eating at Mac's. Find the total. Decide what bills you could use to pay for the food. Figure the change.

Addition and subtraction of money

Keep on Track

Begin the race and continue on the track. Add or subtract the amounts shown in the key.

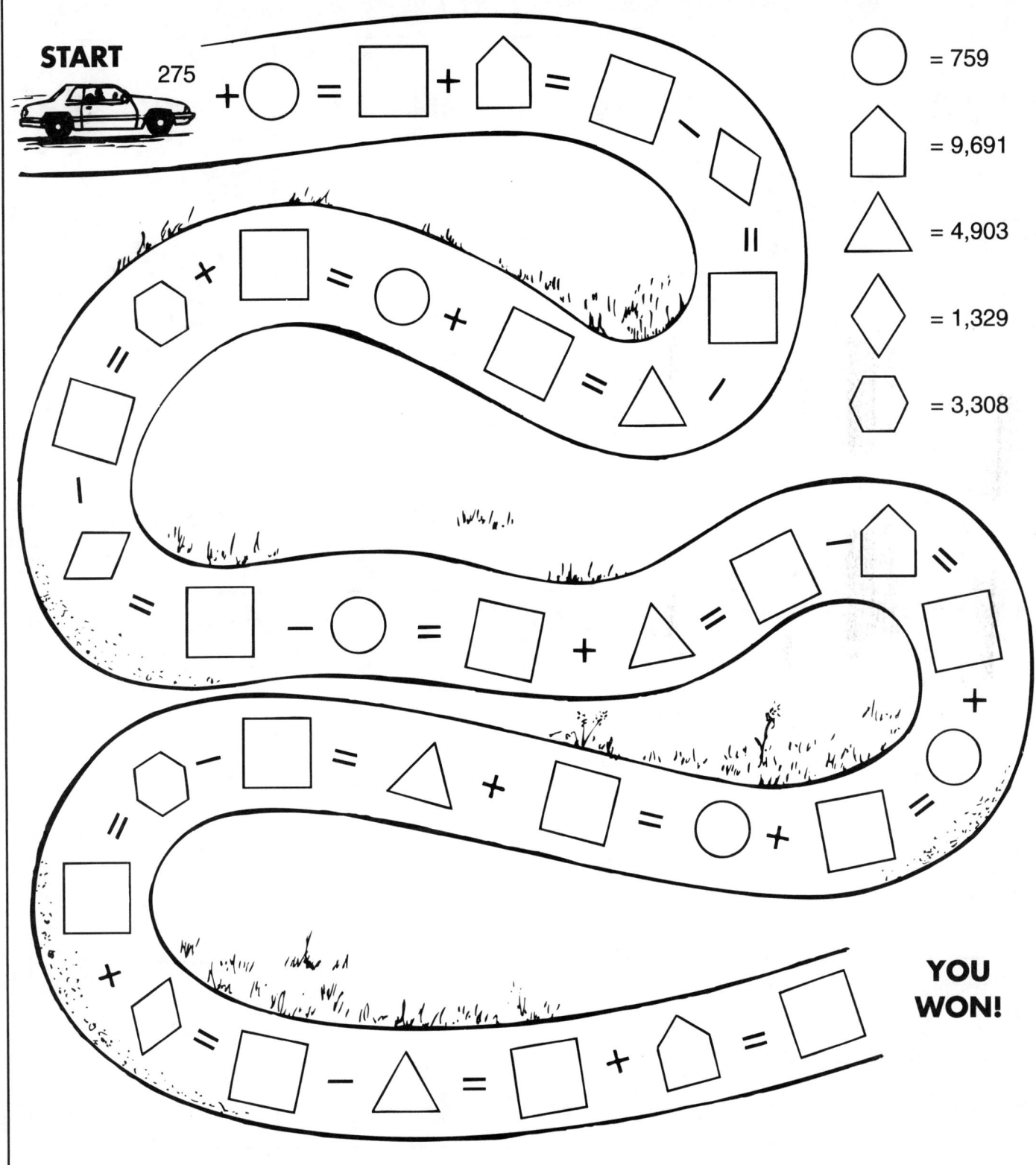

○ = 759
⌂ = 9,691
△ = 4,903
◇ = 1,329
⬡ = 3,308

Addition and subtraction

Name _____

"Sum Difference"

Estimate each sum or difference by rounding each number to the nearest hundred. Add or subtract. Shade in the boxes that have answers of 500 or more.

1. 927 − 419 900 − 400	2. 841 + 219	3. 164 + 99	4. 617 − 302
5. 897 − 143	6. 222 + 170	7. 719 − 504	8. 921 − 647
9. 302 + 256	10. 392 + 184	11. 621 − 304	12. 640 + 599
13. 163 + 173	14. 610 − 601	15. 707 − 290	16. 701 + 291
17. 816 + 219	18. 999 − 702	19. 797 − 142	20. 146 + 208
21. 504 − 392	22. 109 + 99	23. 697 − 64	24. 299 − 199

Estimating sums and differences

Wordy Problems

Solve each problem.

1. On Saturday, 709 people attended a football game. On Sunday, 983 people attended.

   How many more people attended on Sunday? _____

   How many people attended on Saturday and Sunday? _____

2. Barry and Jan were counting stars. Barry counted 179. Jan counted 184.

   How many more did Jan count? _____

   How many did they count altogether? _____

3. Jim bought a toy car for $11.50, a toy bike for $4.73, and $1.28 worth of candy.

   How much money did Jim spend? _____

   If he paid with $20.00, what was his change? _____

4. Mary went to the library to get books on animals and airplanes. The librarian said they had 4,228 books on animals and 3,119 books on airplanes.

   How many more books did the library have on animals? _____

   How many books were there on animals and airplanes? _____

5. While on vacation, Tasha picked out souvenirs for her friends back home. She picked out a hat for $8.26, a T-shirt for $12.79, and a sweatshirt for $17.35.

   How much money did Tasha spend on her friends? _____

   How much more did Tasha spend on the sweatshirt than on the hat? _____

6. The students at Hall Elementary School were competing in a read-a-thon with nearby Johnson Elementary. The students at Hall read 13,986 pages. The students at Johnson read 27,703 pages.

   How many more pages did the students at Johnson Elementary read? _____

   How many total pages did the students at both schools read? _____

   If the fourth-graders at Johnson Elementary read 5,672 pages, how many pages did the other students at Johnson read? _____

Multiplying and Dividing Whole Numbers

INTRODUCING MULTIPLICATION
Literature Activity

Read *Anno's Mysterious Multiplying Jar* by Masaichiro Anno (Philomel Books, 1983). Students will be amazed at how quickly totals multiply into huge, mind-boggling amounts. After reading the book, have students make books or folders and design their own Multiplying Jar books. Share and display the books.

The author's book includes several other multiplying ideas that you can try, such as having the students redesign the classroom using different factors for the number of desks in groups or rows. If possible, try some of the students' arrangements for a few weeks and then discuss which arrangements were best (critical thinking) and which you would like to use for longer periods of time.

DIFFERENT FACTORS, ONE PRODUCT
Manipulative Activity

Using 12 cubes or other counters, have each student or pair of students discover the different ways that 12 can be broken into groups of equal amounts (2 groups of 6, 6 groups of 2, etc.). Students should discover that different numbers can often be divided in different ways. Try other amounts such as 18, 20, 24, and 30.

After students are familiar with dividing objects into even groups and before you teach division with remainders, have them try dividing groups that can't be divided into even groups. For example, students can try dividing 17 into groups of four. They will see that they can make four groups, but one object is left over. It is the remainder. You can begin to show the problems on the board, using R and the number of leftover items for the remainder.

PLAYING CARD MULTIPLICATION
Game

Using a set of 12 playing cards or a set that you make, students in a small group take turns turning over a card. A chosen leader then calls out a number, and the students try to be the first one to say the product of the card turned and the number called out by the leader. In case of a tie, the students who tied have another chance to be first on the next turn. The winners keep the cards and count them at the end of the game.

MULTIPLE FACTS
Class Activity

Given a set of numbers such as 2, 3, and 6, have students write as many facts as possible. You can add extra facts such as 9 and 18 to stretch their thinking.

MISSING FACTS CENTER
Class Activity

Write problems with missing facts on a set of cards. Write the answers on another set of cards. Students can practice by matching and checking their facts. This is a break from the usual flashcard routine.

DRAWING MODELS

Manipulative Activity

Have students draw pictures representing problems. After showing simple facts, such as 3 rows of 5 things for 3 x 5, students can progress to showing larger amounts by drawing place value rods for the tens and adding the ones. For example, three rows of 2 rods and 3 ones will show 23 x 3 = 96.

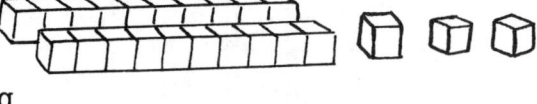
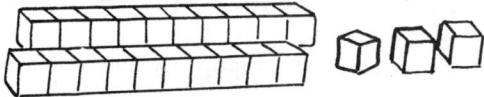
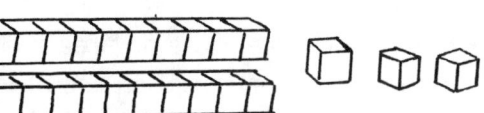

The drawings will be especially valuable when you progress to problems requiring regrouping. Students will see the need to regroup and the need to multiply from right to left. This activity can be altered by using linking cubes instead of drawings to show the factors.

PRACTICE MAKES PERFECT

Mental Math

Have students practice multiplication and division mental math facts 10 minutes each day. Do this by quickly calling out different problems and have students write the answers. Or, write different problems on a piece of paper. Copy the page and give students 10 minutes to complete it. Students need a lot of practice with basic facts so they can say them from memory and not have to depend on a calculator or paper and pencil to figure them out.

DEFINITIONS

Writing Activity

Without using dictionaries, have the students write their own definitions of *multiplication* and *division* in their journals. Although you will find a great variety of answers, you will begin to get insight into the students' thinking. You can have them write the definitions at the beginning of the unit and at the end and compare their answers.

What Can It Be?

Group Activity

Have each student draw or stamp a group of objects on a card. Have students use their cards to ask math riddles. For example, if one student has 6 fish stamped on his or her card, the student can say, "If I multiplied my picture by 8, I'd have 48 fish." This student would not show his or her card with the stamps. Classmates determine the number of objects that are pictured on the card. The student then turns his or her card showing the stamps toward the group to have the other students check the missing fact answer.

REVIEW 0 AND 1 AS FACTORS

Class Demonstration

Some students may not be totally clear on 0 and 1 as factors. For practice, ask 0 x 4 groups of students to come to the front of the room. When no one comes forward, have a student explain why no one came. The student should be able to explain that zero times any number is zero, so no one can come forward. Next, have a student come forward to be a group of one. In another place, have another student be a group of one. Have another student multiply 1 x 2 (or any number of groups you have), showing that 1 times any number is the other number.

FS-23204 Math Made Simple • © Frank Schaffer Publications, Inc.

Geoboard Multiplication and Division

Manipulative Activity

Have students enclose an area such as 5 by 4 on their geoboards. They can then use other colors of rubber bands to divide the area into 4 rows, showing that 20 divided by 4 equals 5. Also, have students start with a row of 3 and make as many groups of 3 as possible. Then they can multiply 3 by the number of groups to find the product. Several students can work together with the boards to involve higher numbers.

MULTIPLICATION IS COMMUTATIVE

Manipulative Activity

Remind students that they discovered that addition was commutative and that subtraction was not. Have them discover if multiplication is commutative.

Have students link 4 sets of 6 cubes showing the sentence 4 x 6 = 24. Next, have them link 6 sets of 4 cubes each (6 x 4) and multiply. They will see that the answer is the same either way and that multiplication, as well as addition, is commutative. Have students try division to see that it is not commutative.

FINDING AVERAGES

Class Activity

To help students learn about averages, begin with simple problems such as "How many pets do you have?" After adding the number of pets students in your room have, divide this by the number of students in the class. The quotient will be the average number of pets per student. Discuss that this average includes some students who have several and some who may have none. Students will begin to see that averages represent a group rather than showing what each person has.

SPINNER FUN

Game

Divide students into small groups. Each student in the group takes turns spinning a spinner two times. The students multiply the two numbers. On a piece of paper, the student who did the spinning draws the facts to show his or her multiplication problem and writes the number sentence. Play continues in the same manner. Each player in the group gets 8 turns. At the end of the game, each player adds all 8 answers together. Players compare with other group members. The player with the largest total wins. If necessary, the players can use calculators to find their totals.

To change this to a division game, a player can spin twice for the dividend number and once for the divisor. The player then divides and records the quotient in picture-form with the number sentence under it. If there is a remainder, the player can add ½ a point to the total when adding the quotients for a final score, or the remainders can just be dropped.

EVERYDAY MULTIPLICATION AND DIVISION *Class Activity*

Have the students find the number of days in 6 weeks (7 x 6) or the number of weeks in 56 days (56 ÷ 7). You and the students will think of a lot of daily ideas such as the number of eggs in five dozen or the number of dozens that can be made from 36 eggs.

LEARN TO DIVIDE *Manipulative Activity*

Have students divide groups of cubes into smaller equal groups, showing the leftovers as remainders. Next, write the number of students in your class on the board. At a given signal (a bell, clap, etc.), have students divide themselves into groups of 2. On a chart, write the sentence 30 (or the number of students in your class) ÷ 2 = 15. If you have an uneven number of students, you can be a member so it will divide evenly into 2 groups, or you can show a remainder. Repeat the signal, having students divide into groups of three. On the chart, show 30 ÷ 3 = 10. Proceed to groups of 4, 5, 6, 7, 8, and 9, adding to the chart as you go along. Use this as your introduction to division with and without remainders. Students can continue to use counters to divide, but they should not rely on the counters too long as working with large numbers of counters is not always possible or a good idea.

LARGEST AND SMALLEST QUOTIENTS *Class Activity*

Write 3 (or more) numbers on the board. Have the students decide how to place the numbers so that they will get the largest quotient when they divide. After experimenting, have them come up with guidelines. They may want to use the smallest possible number as the dividend. For example, if 3, 6, and 8 were placed on the board, students may set up the equation 86 ÷ 3. Play this as a small team game giving points to all the teams that have found the largest quotient. Another time, reverse the goal and have students try to find the smallest quotient, a quotient without a remainder, etc. You can put more numbers on the board for the students, and also tell the students that they do not need to use all numbers listed.

School-Home Connection *Homework*

 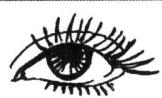
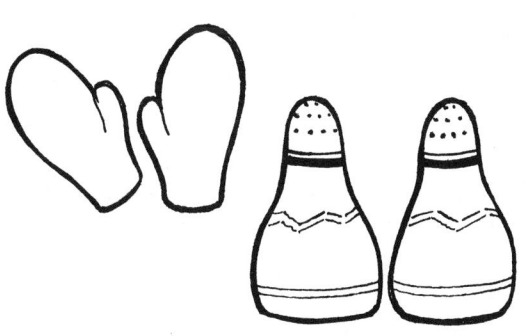

As a homework assignment, have students think of all the things they can that come in twos. It's easy to start with the body—arms, legs, eyes, etc. Then students can move on to things like twins, members of a twosome or duet, etc. After the students bring in their "twos," make a chart and display it in the room. Use it for extending students' thinking. For example, ask students such questions as "If one person has two legs, how many legs will six people have?"

As you review each set of facts, have the students make lists of things that come in threes, fours, fives, etc. Make new charts and think of new problems. Students should be thinking of problems to share with the group as often as possible.

Name_____

Multiplying All Around

Multiply the center numbers by the numbers in the middle rings.

1.
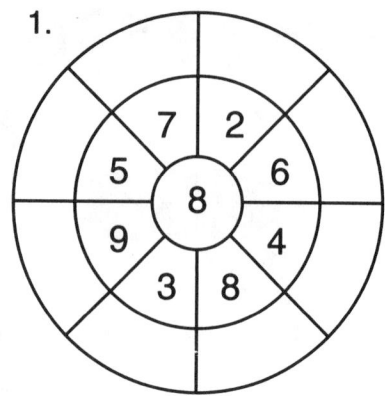

2.
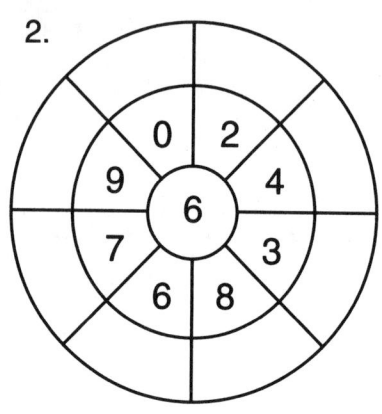

3.
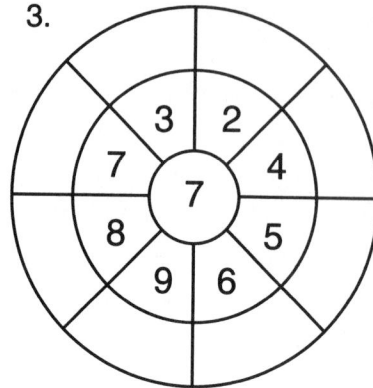

4.
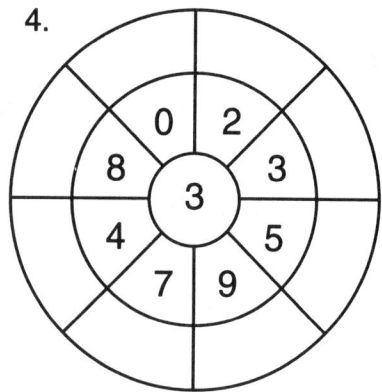

5.
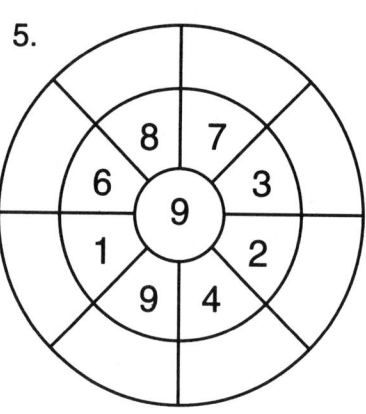

6.
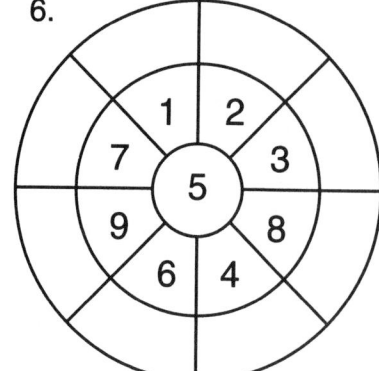

7.
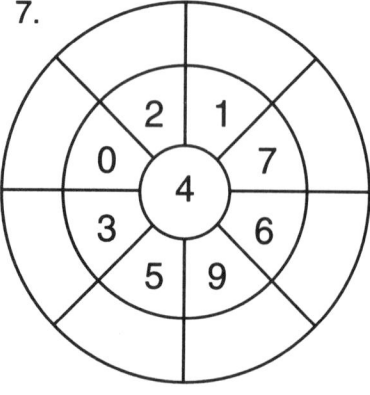

8.
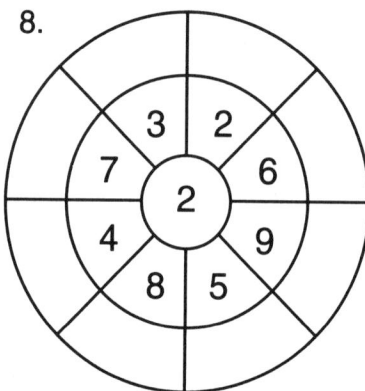

9.
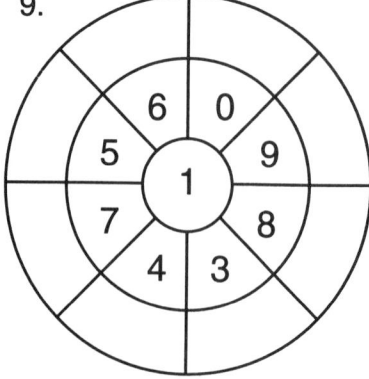

24 Multiplying 1–9 as factors

FS-23204 Math Made Simple • © Frank Schaffer Publications, Inc.

Color Crazy

Multiply. Color the products.

blue = 400–499   yellow = 300–399   green = less than 300

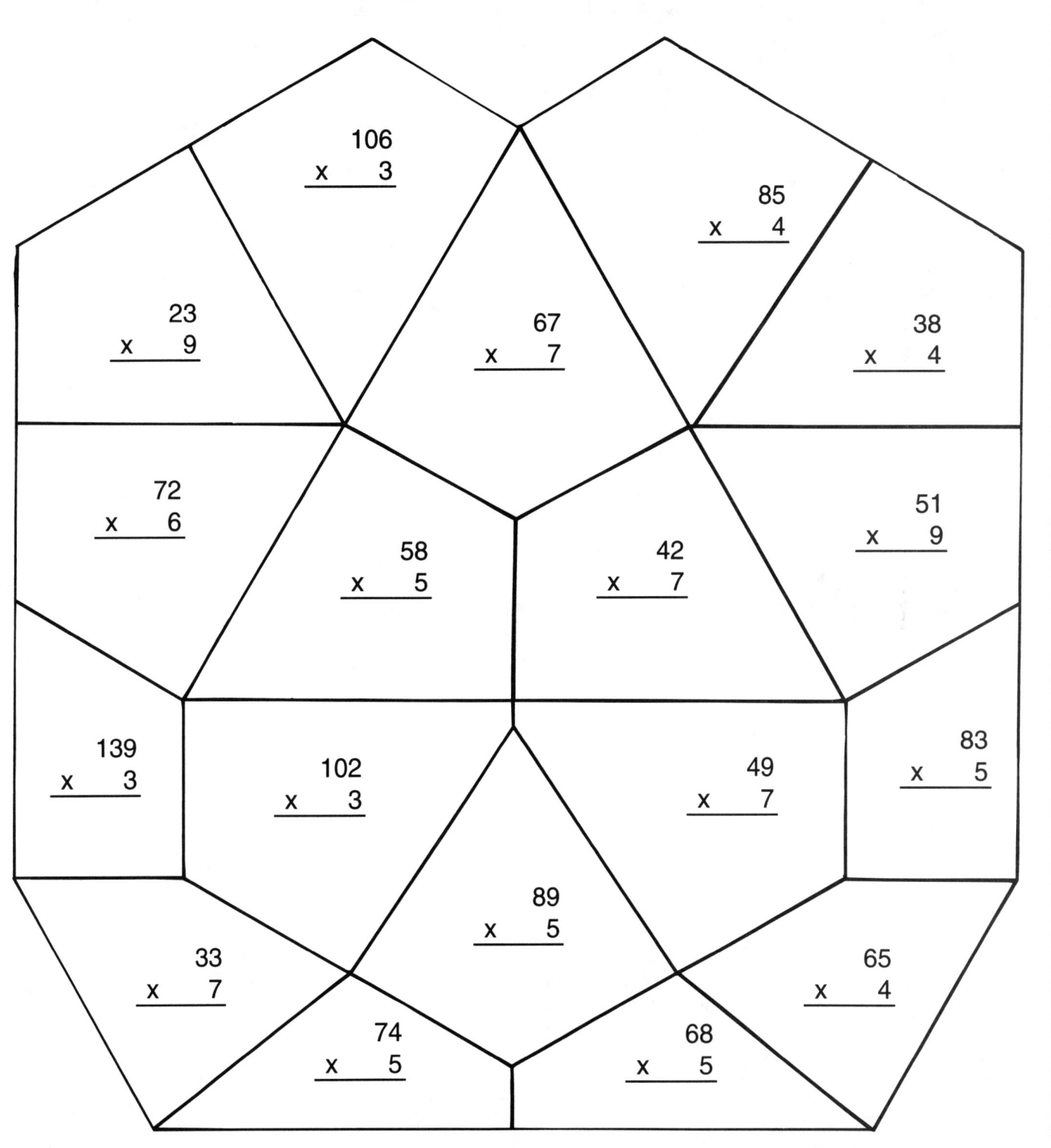

Multiplying by 1-digit numbers

Get in Shape

Multiply.

A.
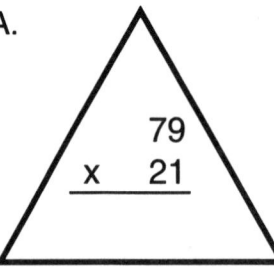
      34
x   67

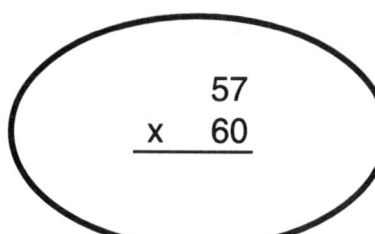

      29
x   16

B.
   603
x   19

   101
x   38

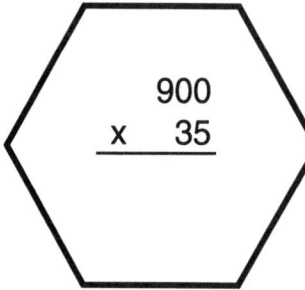

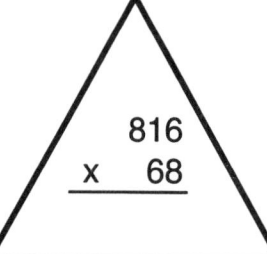

C.
   144
x   30

   175
x   14

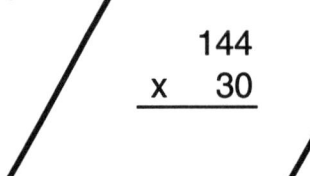

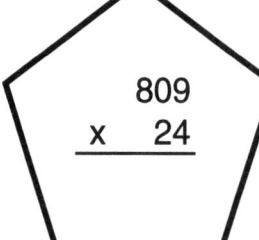

D.
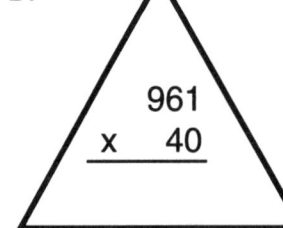

   506
x   12

    56
x   22

    63
x   18

Up, Up, and Away

Multiply. Put the letters of the answers on the lines below to find out a secret message.

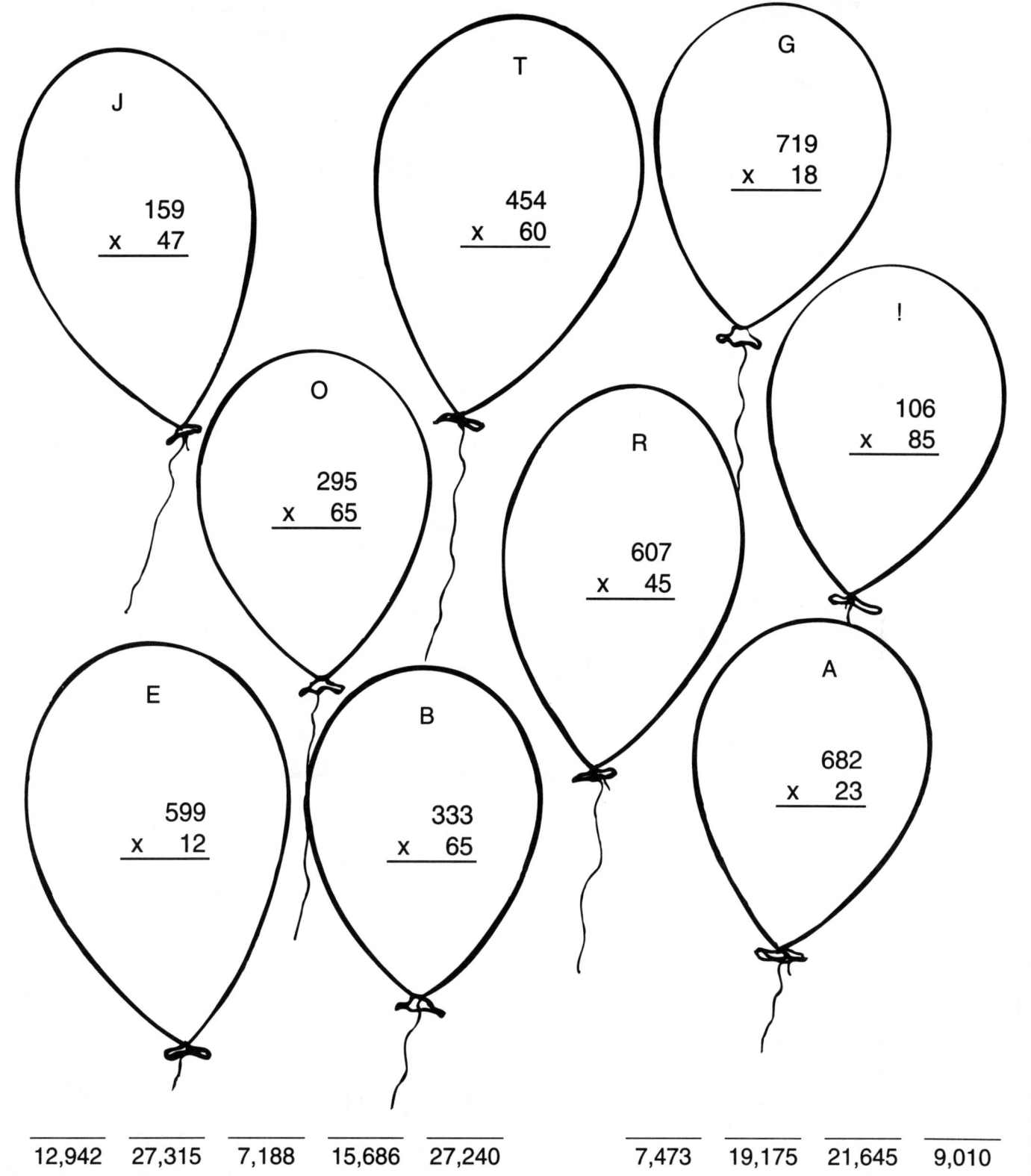

12,942   27,315   7,188   15,686   27,240   _____   7,473   19,175   21,645   9,010

Multiplying by 2-digit numbers

Moving Along

Divide each problem. Write the quotient as the divisor for the next problem. Write each final answer in a circle.

1. 5)25 → 5)40   )56   )63   )54   ◯
2. 9)18   )18   )81   )72   )64   ◯
3. 4)36   )27   )21   )63   )36   ◯
4. 5)40   )16   )16   )24   )18   ◯
5. 6)30   )45   )45   )10   )8   ◯
6. 7)35   )20   )32   )24   )21   ◯
7. 2)4   )10   )15   )18   )30   ◯
8. 5)35   )14   )16   )64   )32   ◯

Shade in the answers that are correct.

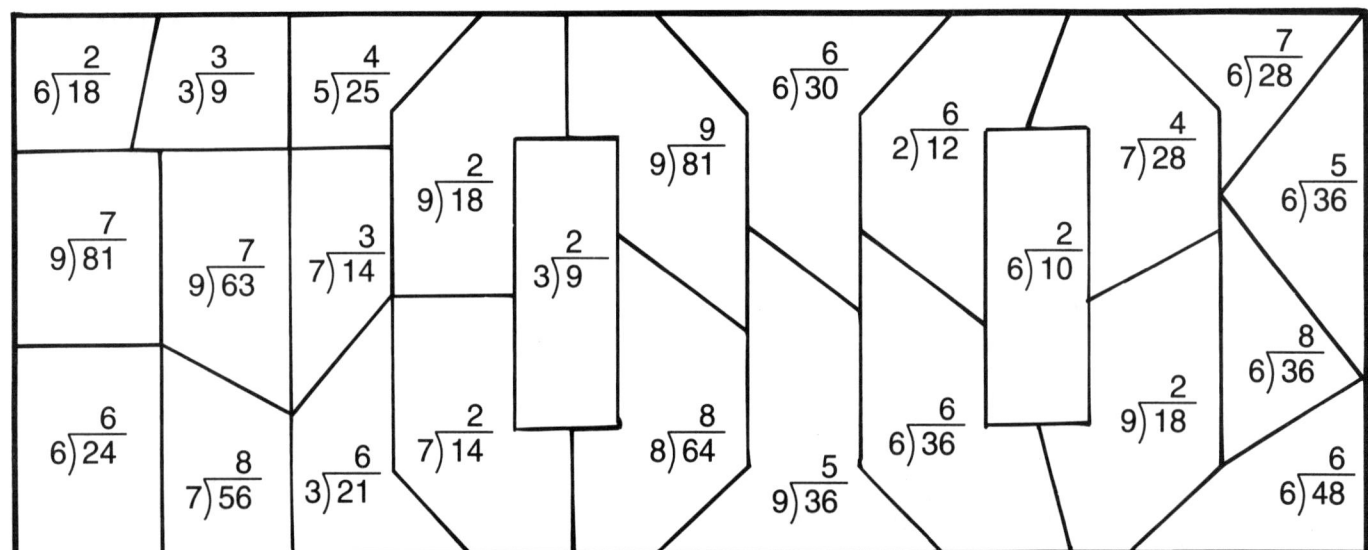

28 Dividing 1–9   FS-23204 Math Made Simple • © Frank Schaffer Publications, Inc.

A-Mazing Division

Divide. Check by multiplying.

1. 8)480   60 / 60 × 8
2. 9)369
3. 5)250
4. 7)560
5. 6)240
6. 4)364

Divide. Watch out for remainders.

7. 9)60
8. 8)47
9. 8)75
10. 4)31
11. 3)23
12. 8)47

Follow your answers through the maze.

Start

60	41	50	80	30	70	91	13	14
40	4R6	5R1	40	91	6R6	8R2	9R1	3R2
70	18	24	6R1	4R2	5R7	7R4	8R2	9R3
5R5	7R3	6R5	5R2	9R1	9R3	7R3	7R2	5R7

You did it!

1- and 2-digit quotients with and without remainders

Banana Boat

Divide and write the quotients in the puzzle. Hint: R for remainder will take up its own space.

ACROSS

1. 6)789
4. 6)332
6. 4)291
8. 9)189
10. 7)200

12. 8)651
13. 2)827
15. 6)222
16. 5)396

DOWN

2. 6)188
3. 8)298
4. 4)212
5. 9)244
7. 8)511

9. 7)708
11. 9)394
14. 4)432

Zero In

Divide. Watch out for the zeros.

1. 3)914

2. 7)765

3. 8)6472

4. 6)609

5. 5)800

6. 9)975

Write the quotients above in order from smallest to largest.

1. _____ 2. _____ 3. _____ 4. _____ 5. _____ 6. _____

Write the missing digits in each problem.

7. 5☐ R2
    5)252
    2☐
    ☐

8. 83☐
    6)4980
    4☐
    1☐
    18
    ☐

9. 5☐ R☐
    4)203
    ☐☐
    ☐

10. 9☐9
    9)8181
    ☐☐
    ☐1
    ☐1
    0

11. 8☐
    3)261
    ☐☐
    ☐☐
    21
    ☐

12. 5☐1 R☐
    5)2☐09
    2☐
    0☐
    5
    ☐

13. ☐☐
    8)656
    64
    16
    ☐☐
    ☐

14. 2☐☐
    7)1435
    ☐☐
    ☐☐
    ☐☐
    ☐

FS-23204 Math Made Simple • © Frank Schaffer Publications, Inc.

Zeros in quotients 31

Name_____

MESSAGE

Divide It Up

Divide. Write the letter of each problem above its answer below.

I.	I.	O.	D.
34) 2584	42) 8526	63) 6237	55) 3575

D.	Y.	T.	U.
89) 6319	90) 1800	31) 6231	18) 882

___ ___ ___   ___ ___ ___   ___ ___ !
20   99   49   65   203   71   76   201

Find the total test scores below. Then find the averages.

Students	Test Scores					Total	Average
Sue Ann	18	20	17	19	16		
James	15	14	17	16	13		
Tommy	19	19	20	18	19		
Carlos	16	19	18	17	16		
Anna	20	20	19	20	19		

Who has the highest average? _____

Who needs to study more? _____

Dividing by 2-digit numbers

Doggie Trouble

Help the dog get to his bone by connecting the correctly-done division problems to make a path.

23)1972 85 R17	32)989 30 R29	43)629 14 R27	30)6963 232 R3	65)6959 107 R4
15)1508 104 R8	22)687 31 R4	19)582 31 R12	87)6933 78 R6	52)8071 155 R11
31)8013 255 R15	42)6231 14 R15	26)541 20 R18	18)819 54 R9	26)809 31 R3
60)3207 52 R27	39)6193 158 R31	27)983 36 R11	71)8203 115 R8	17)989 58 R3
		81)1484 18 R26	44)1943 44 R7	28)1773 63 R9

Dividing by 2-digit numbers with remainders

Decimals and Fractions

In this section, students will relate decimals to tenths and hundredths by means of many stimulating and challenging activities. Relating decimals to money provides a concrete understanding for most students and establishes a vital connection between math and daily life.

Also by completing the activities in this section, students will see fractions as naming equivalent parts of a whole or a group. They should use manipulatives to compare fractions and mixed numbers. Manipulatives can be used to help students develop a concrete understanding of addition and subtraction of like and unlike fractions, before moving on to written symbols.

CONCEPTS

The ideas and activities presented in this section will help students explore the following concepts:
- relating fractions and decimals
- writing and comparing decimals
- adding and subtracting decimals
- multiplying and dividing by multiples of 10
- greatest common factors
- least common multiples
- reducing fractions
- equivalent fractions
- improper fractions and mixed numbers
- comparing fractions
- adding and subtracting fractions with like and unlike denominators

USING PLACE VALUE BLOCKS *(Manipulative Activity)*

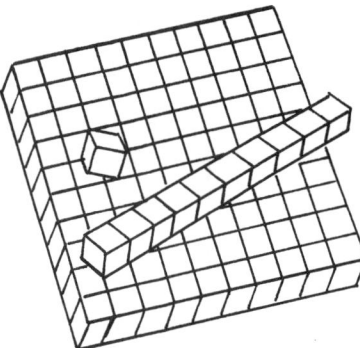

Place value blocks are excellent manipulatives for decimals. Each pair of students should have 10 ones, 10 rods (10), and a flat (100). Using the ones, students can show that one is equal to 0.1, or ¹⁄₁₀, of the 10 rod or 0.01, or ¹⁄₁₀₀, of the 100 flat. Using various amounts, such as 4 ones blocks, students can write the decimal 0.4 and then draw a square to represent each block. Have them repeat the process with various amounts. They can show the ones blocks as hundredths of the hundred flat, writing 0.04 for the four ones as part of 100. Drawing pictures reinforces the concept visually and can be referred to when the blocks are put away.

UNDERSTANDING THE VALUE OF DECIMAL PLACES *(Class Demonstration)*

A common thinking error with students is that longer numbers are larger. Show students 0.3 on the board. Extend the 0.3 to show that it is the same as 0.30. Next, write 0.29. Use graphs of 100 blocks to color in 30 (0.3) and 29 (0.29), showing that even though 0.29 may have looked larger at first glance, it is actually smaller than 0.3.

SMALLER AND LARGER *(Writing Activity)*

To be sure that students understand that as the denominator gets larger, the pieces get smaller, have them compare a ⅙ piece of pie graph with a ⅓ piece of pie graph. In their math journals, have students write a paragraph explaining this idea—that the larger the denominator, the smaller the piece. If denominators are the same, then the numerator determines which amount is greater. Have students lay ⅗ over ⅖ to see this concept clearly. They should draw pictures showing these ideas in their journals.

MORE THAN, LESS THAN, EQUAL TO RESPONSES

Class Activity

Give each student three different-colored pieces of paper. Using all three colors, have all students use the same color of paper on which to write a greater than sign, another color for a less than sign, and another color for an equals sign. Then write two fractions. Have the students hold up the correct signs for comparison of the fractions. Since students will be using the same colors of paper, you will be able to see who needs help easily.

Another version of this is to prepare a variety of fraction cards. Divide the students into groups and have one student in each group draw two fraction cards. The other students hold up a <, >, or = card to compare the fractions.

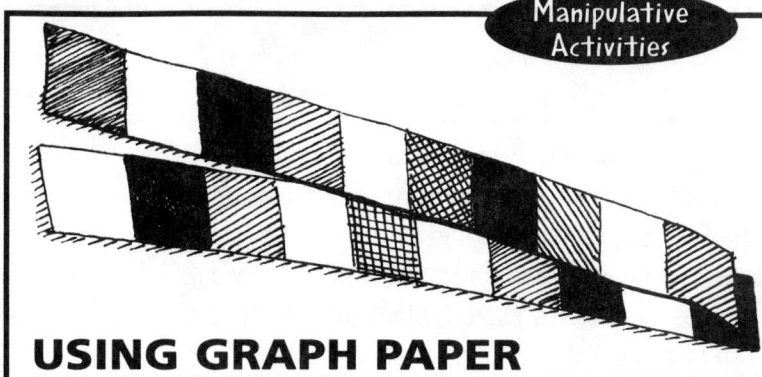
Manipulative Activities

USING GRAPH PAPER

- Students can cut out a row of 10 squares from graph paper. Or, they can use this as a measurement exercise by drawing a bar, marking it in 10 equal segments, and cutting it out. Have students write ¹⁄₁₀ on each segment. Have students use the strips to compare fractions or decimal amounts.

- Using a sheet of graph paper, have each student outline a group of 10 squares and then color in a given decimal amount. For mixed numbers, students can outline and color 10 squares for each whole number, plus the sections of 10 that represent the fraction.

STORY PROBLEMS

Writing Activity

Have the students write story problems that involve 100 things (decimals in the hundredths place), or 10 things (decimals in the tenths place). For example, "If there are 100 seats in the school auditorium and 82 children attend the class play, what fraction of the seats are occupied? (⁸²⁄₁₀₀ = 0.82)," or "If 10 children usually practice with the soccer team, but 3 were sick, how many went to practice? (⁷⁄₁₀ = 0.7) How many were sick? (³⁄₁₀ = 0.3)"

Art Project

Clay Shapes

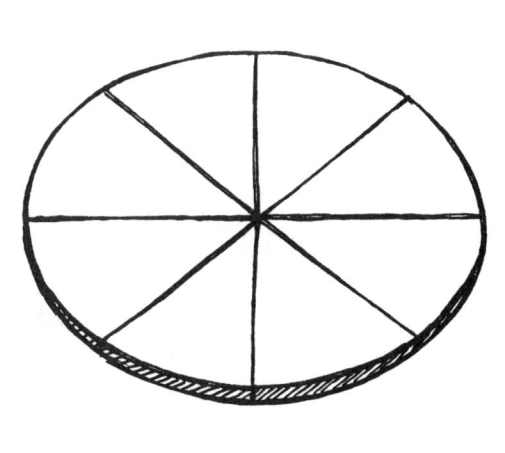

A very tactile experience could involve having students roll out clay, cut it into circles, and cut the clay "pies" into fractions. These pie pieces could then be painted and fired if desired. If different students cut different fractions for their pies and use different colors, a center activity could be created involving a lot of questions for comparing and matching. You can write on the fired clay with a permanent marker, so the amounts could be shown if desired. If you write the amounts on the bottoms of the pies, students could put the pieces together and determine the fractional parts and then check the answers.

FRACTION PICTURES

Art Project

Have the students draw pictures using given fractions. For example, ½ of their pictures must be made using trapezoids and ¼ using triangles. The other ¼ may be made from students' choices of shapes. Be sure to make it clear that the fraction refers to the total number of shapes used.

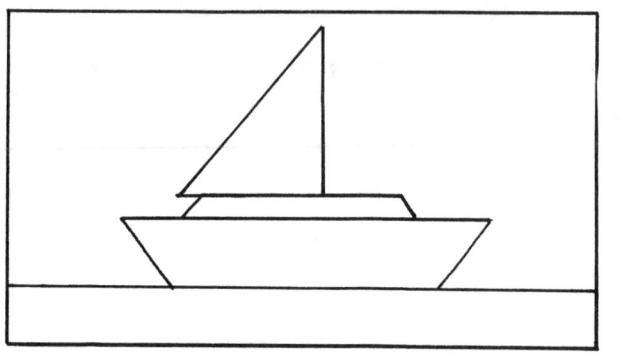

SHOWING AMOUNTS IN DIFFERENT WAYS

Group Activity

Write different amounts on cards in one of three ways (³⁄₁₀, 0.3, three tenths). Use large cards for group activities and small cards for small group or center activities. Have the students use the cards to match like amounts and to put the numbers in size order, largest to smallest or smallest to largest. Keep repeating this type of activity until you are sure that students understand that 4.9 is smaller than 26, etc. This may be difficult for some students until they have worked with the manipulatives.

HUMAN FRACTIONS

Class Demonstration

Using the total number of students in your class as the denominator, have students write fractions representing the following: themselves (¹⁄₃₁—or whatever applies to your class), the girls (¹⁴⁄₃₁), the number of boys (¹⁷⁄₃₁), the number with green eyes, etc.

Have a small group of students come forward. Give suggestions for different attributes students can express in fractions. For example, in a group of 7, there may be ³⁄₇ boys, or ⁶⁄₇ students with brown hair, etc. The possibilities are endless, and this concept will stimulate creative thinking if you encourage students to find one more way to tell about this group. Switch to different groups so that everyone is in a group at some time.

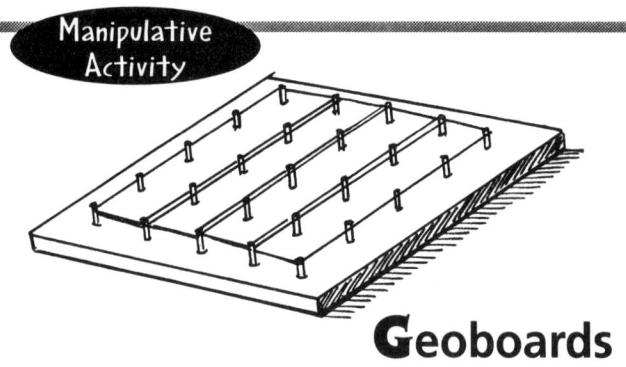

Geoboards

Manipulative Activity

Have students use rubber bands to divide their geoboards into different fractional parts. It is good to have students transfer their fractions onto geoboard dot pages and write the fractions and color or outline the parts. Have them experiment, trying to divide the boards into fourths, fifths, etc., by trying different shapes.

VOCABULARY

Group Activity

Have the students add *fractions, numerator, denominator, equivalent fractions,* and *lowest terms* to their math journals.

MISSING DECIMALS

Group Activity

On the front of a card, write a decimal. It can be tenths or hundredths. On the back, write the amount needed to total 1 whole. (For example: Front 0.7; Back 0.3) Divide students into groups. A student in each group draws a card, reads the front, and asks another member of the group to determine the amount needed to make one whole. He or she checks the answer on the back, and the responding student chooses and reads the next card.

DAY-TO-DAY FRACTIONS

Class Activity

As you take roll, have the students determine what fraction of the group is present, what fraction is wearing something red, etc.

Students can also determine what part of the month has passed, such as 5/31 days. Sometimes students will be able to reduce the fractions to lowest terms.

LOWEST TERMS

Class Activity

Fractions such as ½, ⅚, and 9/10 are in lowest terms because both the numerator and denominator can only be divided by one. Fractions such as 4/6 and 5/10 can be reduced to lowest terms by dividing both the numerator and the denominator by the same number. Both the 4 and the 6 can be divided by 2 resulting in ⅔. The ⅔ is equivalent to 4/6, but the ⅔ is in lowest terms.

Let students use graph paper or fraction strips to show equivalent fractions. Students who know their division and multiplication facts will have a much easier time with reducing fractions.

School-Home Connection

Homework

Have students brainstorm with their parents and list things that are commonly cut into fraction parts, such as sandwiches (halves or fourths), pies, (sixths or eighths), etc. They may ask restaurants how many pieces their cakes and pies are divided into. They can then find out the price per piece (divide the cost by the number of slices), and see if buying the whole item is a better buy. Have parents also help their children figure out how much is saved when the whole item is bought instead of pieces. In their journals, students could discuss this idea as well as which type of dessert they prefer and why.

DECIMALS AND FRACTIONS

Match Ups

Decimals can be used to show tenths. (0.6 = $\frac{6}{10}$) Write the decimals for the shaded parts.

 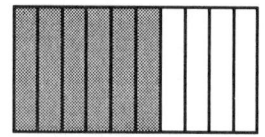 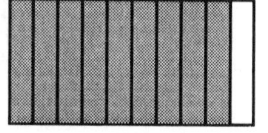 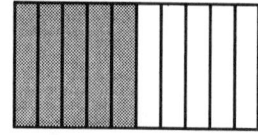

1. _____ 2. _____ 3. _____ 4. _____

5. Match the decimals with the fractions and word names.

$\frac{8}{10}$	0.4	seven tenths
$\frac{1}{10}$	0.8	four tenths
$\frac{7}{10}$	0.1	eight tenths
$\frac{4}{10}$	0.7	one tenth

Whole numbers are shown before the decimal. Example: two and nine tenths = 2.9
Write the whole numbers and decimals for the shaded figures.

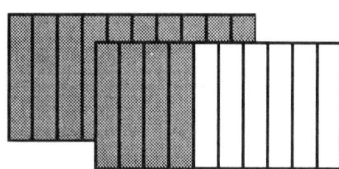 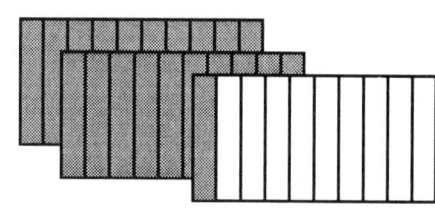 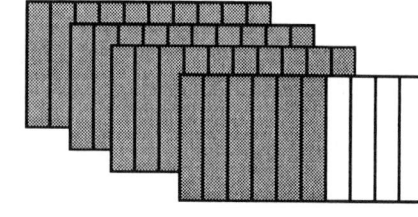

6. _____ 7. _____ 8. _____

Hundredths can be shown using decimals, too. twelve hundredths = 0.12

9. Match the decimals for the shaded parts.

4.31

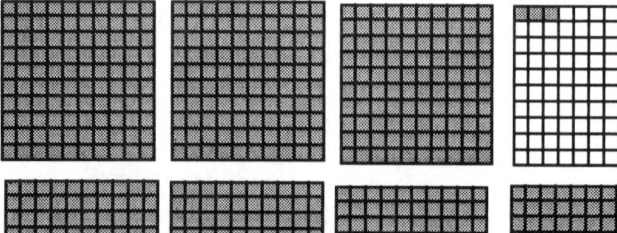

1.22

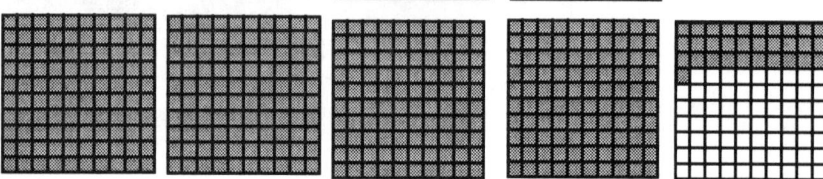

3.03

Decimal Dilemma

Write the decimals two other ways.

2.06	$2\frac{6}{100}$	two and six hundredths
6.9		
4.73		
0.84		
7.4		

Circle the largest decimal in each section and draw a box around the smallest decimal. If the decimal circled is in the middle of the group, color the section red. If it is on the right, color it blue. If it is on the left, leave it white.

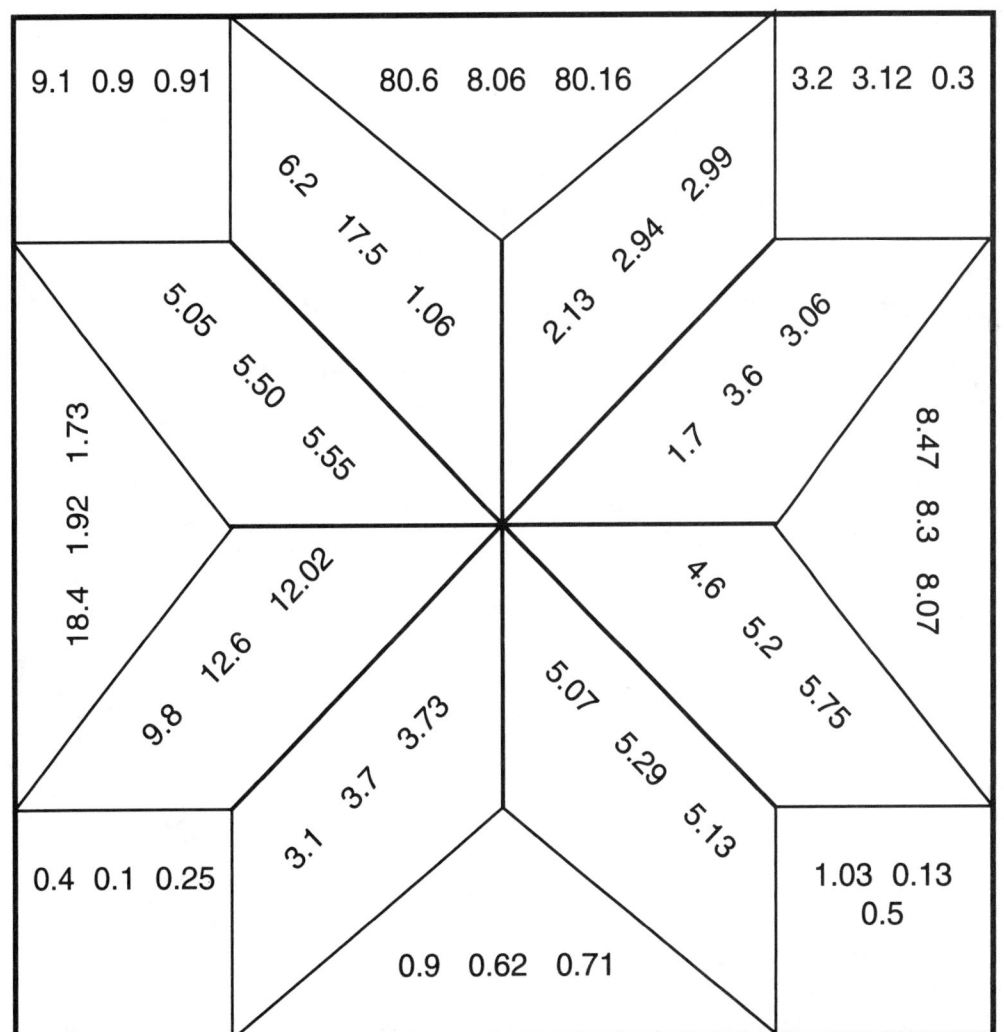

Writing and comparing decimals

What's Missing?

Add or subtract.

A.
```
   0.9        5.7        6.04       5.55      10.10       11.7
+  4.6     −  2.9     +  3.12    −  4.60    −  4.65    +  10.1
```

B.
```
   9.1       51.6      70.99      18.4        9.99       6.71
   7.5       27.9      61.73      20.7        2.73      12.58
+  6.2    + 18.8    + 47.12    + 32.5     +  1.84    +  9.01
```

C. 11.07 + 9.91 =   27.9 − 12.6 =   13.21 − 7.54 =

D. Write the missing digits.

```
   4.□6        □.46        8.□□       92.7       □□.9
+  5.7□     −  3.□5     +  5.66     − 6□.9    −  38.□
  1□.10        3.5□       1□.10       □0.□       41.6
```

E. 5.66 + □.□□ = 9.99   12.6 + □□.□ = 22.9

  □.□ − 3.17 = 3.63

F. Shade in the boxes where the decimals are in numerical order.

4.9, 6.2	1.7, 1.6	12.06, 13.2
12.4, 9.6	8.0, 8.01	2.7, 2.67
5.01, 5.35	2.66, 2.65	72.9, 79.2
2.13, 2.03	1.6, 1.7	5.00, 4.9
2.04, 2.14	4.01, 4.00	18.6, 18.65
3.1, 3.01	9.9, 10.01	2.75, 2.70

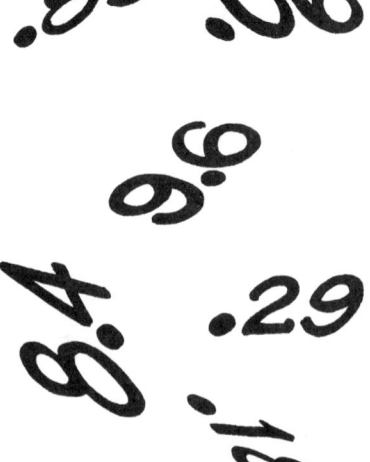

Adding and subtracting decimals; comparing decimals

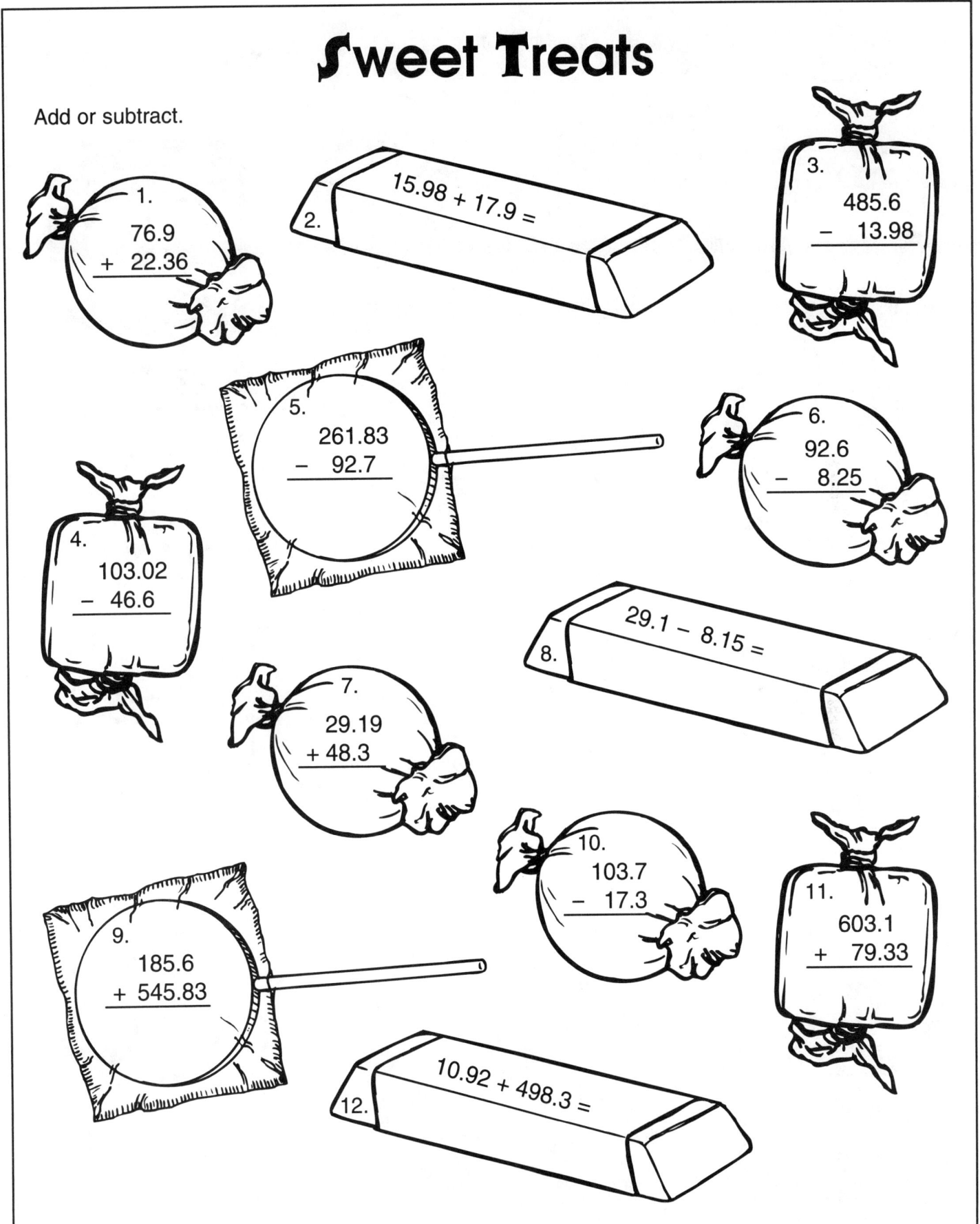

Drummin' Up Answers

Multiply or divide. Write the answers in the drums.

1. 3.8 x 10

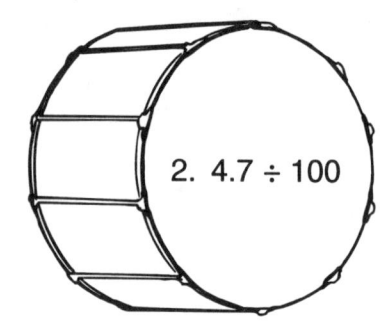
2. 4.7 ÷ 100

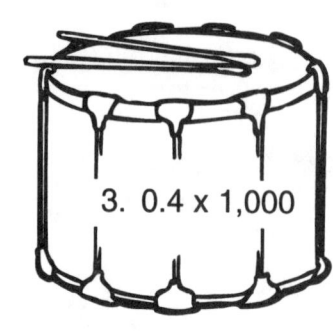
3. 0.4 x 1,000

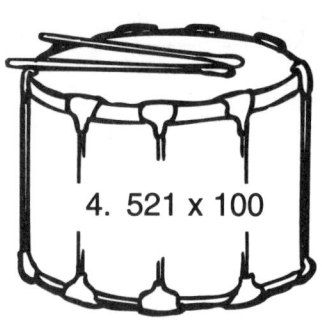
4. 521 x 100

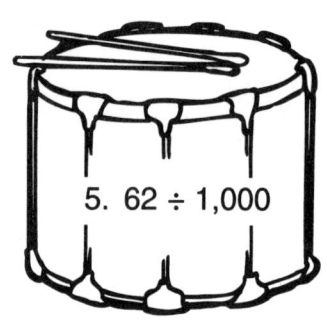
5. 62 ÷ 1,000

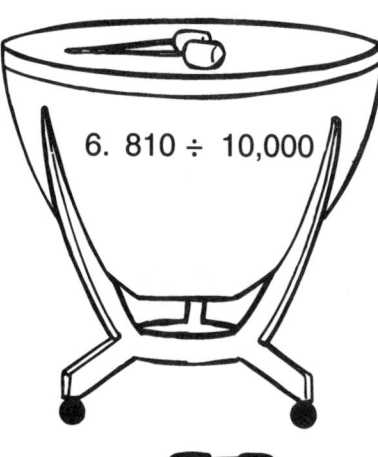
6. 810 ÷ 10,000

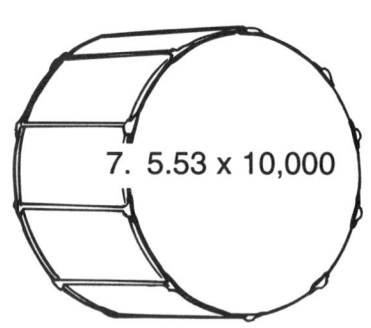
7. 5.53 x 10,000

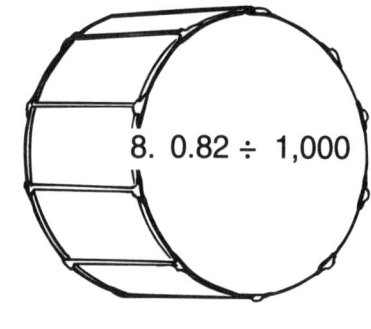
8. 0.82 ÷ 1,000

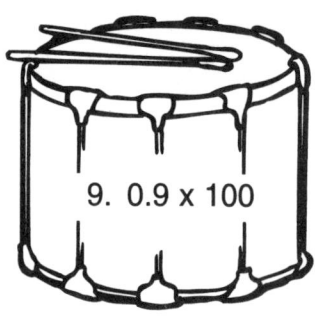
9. 0.9 x 100

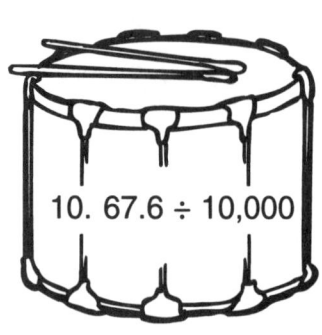
10. 67.6 ÷ 10,000

11. 33.7 x 1,000

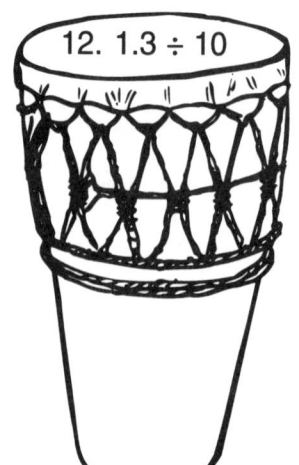
12. 1.3 ÷ 10

Multiplying and dividing by multiples of 10

Sports Smart

Draw lines from the numbers in the tennis rackets to their greatest common factors in the balls.

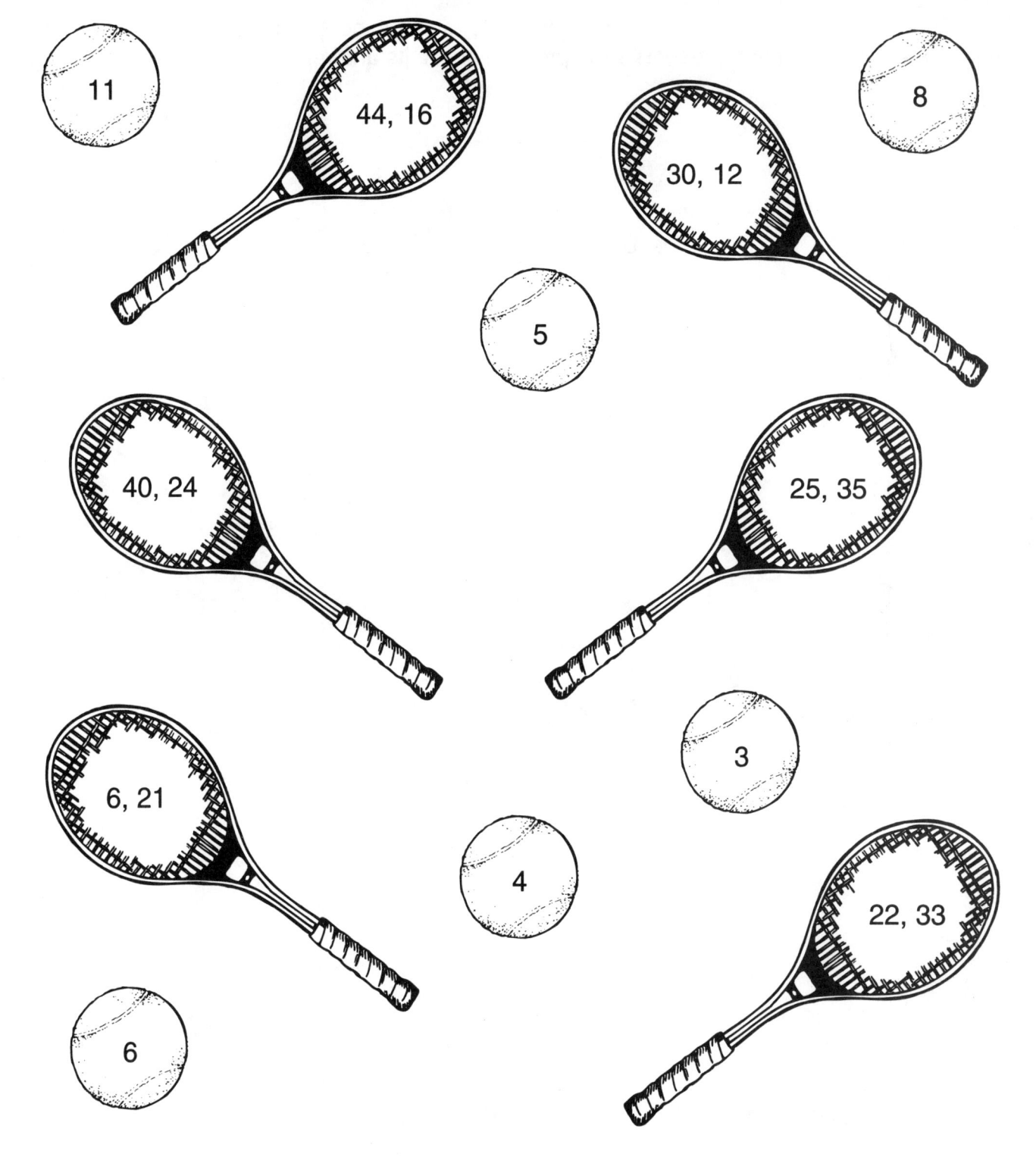

Greatest common factors

Swampy Math

What is the name of the biggest swamp in the world?

To find out, find the least common multiple for the pairs of numbers below at the bottom of the page. Write the letter that represents each problem above its answer.

G.	18, 21	N.	6, 21	T.	22, 6	N.	25, 30
R.	8, 6	A.	16, 6	A.	9, 15	L.	9, 30
A.	9, 12	A.	12, 21	D.	8, 18	N.	14, 4
P.	25, 35						

___ ___ ___ ___ ___   ___ ___ ___ ___ ___ ___ ___ ___
126 24 36 150 72   175 45 42 66 48 28 84 90

Name _____

Snowflake Fun!

Reduce the fractions below and write them in word form.

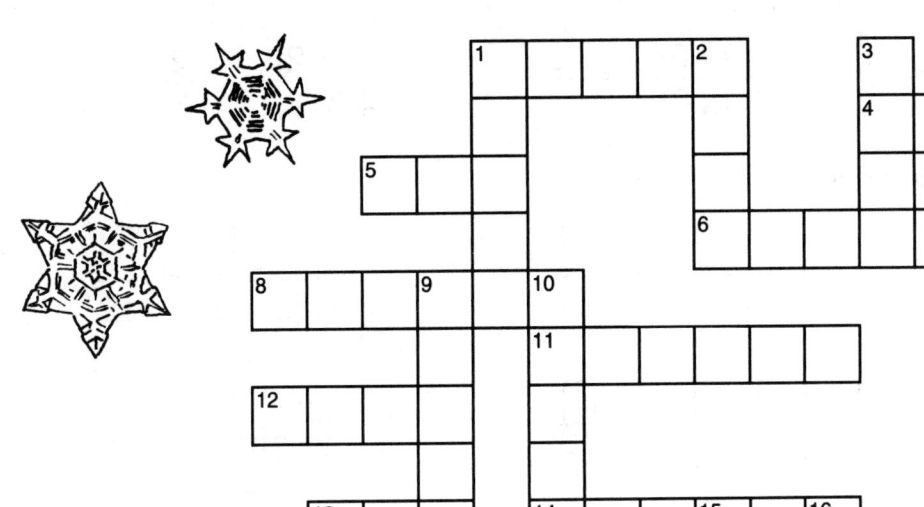
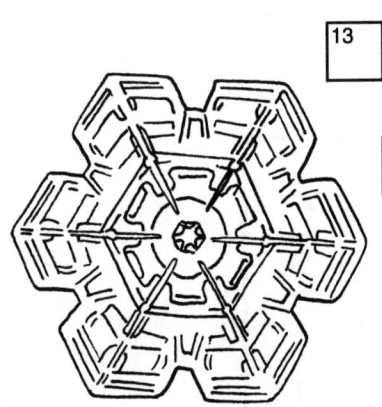
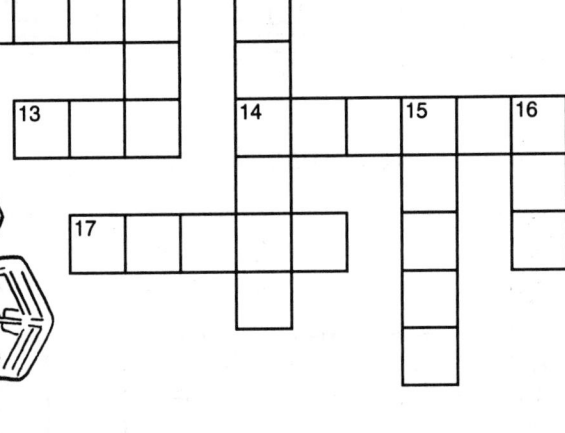

ACROSS

1. $\frac{4}{36}$ = one _____

4. $\frac{4}{28}$ = _____ seventh

5. $\frac{30}{33}$ = _____ elevenths

6. $\frac{9}{12}$ = three _____

8. $\frac{15}{18}$ = five _____

11. $\frac{7}{56}$ = one _____

12. $\frac{16}{36}$ = _____ ninths

13. $\frac{9}{72}$ = _____ eighth

14. $\frac{25}{45}$ = five _____

17. $\frac{24}{33}$ = _____ elevenths

DOWN

1. $\frac{3}{27}$ = one _____

2. $\frac{9}{18}$ = one _____

3. $\frac{20}{35}$ = _____ sevenths

7. $\frac{2}{4}$ = one _____

9. $\frac{9}{21}$ = _____ sevenths

10. $\frac{12}{42}$ = two _____

15. $\frac{8}{24}$ = one _____

16. $\frac{12}{26}$ = _____ thirteenths

Reducing fractions

Name_____

Equal Time

To find equivalent fractions, the numerator and denominator are multiplied by the same number.

$$\frac{1\,(\times 3)}{3\,(\times 3)} = \frac{3}{9} \qquad \frac{1\,(\times 6)}{2\,(\times 6)} = \frac{6}{12} \qquad \frac{1\,(\times 2)}{4\,(\times 2)} = \frac{2}{8}$$

Change each fraction to higher terms.

1. $\dfrac{1\,(\times \square)}{5\,(\times \square)} = \dfrac{\square}{10}$
2. $\dfrac{2\,(\times \square)}{5\,(\times \square)} = \dfrac{\square}{10}$
3. $\dfrac{1\,(\times \square)}{6\,(\times \square)} = \dfrac{\square}{12}$

4. $\dfrac{2\,(\times \square)}{7\,(\times \square)} = \dfrac{\square}{21}$
5. $\dfrac{1\,(\times \square)}{2\,(\times \square)} = \dfrac{\square}{12}$
6. $\dfrac{2\,(\times \square)}{3\,(\times \square)} = \dfrac{\square}{12}$

Multiply in your head. If the equivalents are correct, shade in the areas.

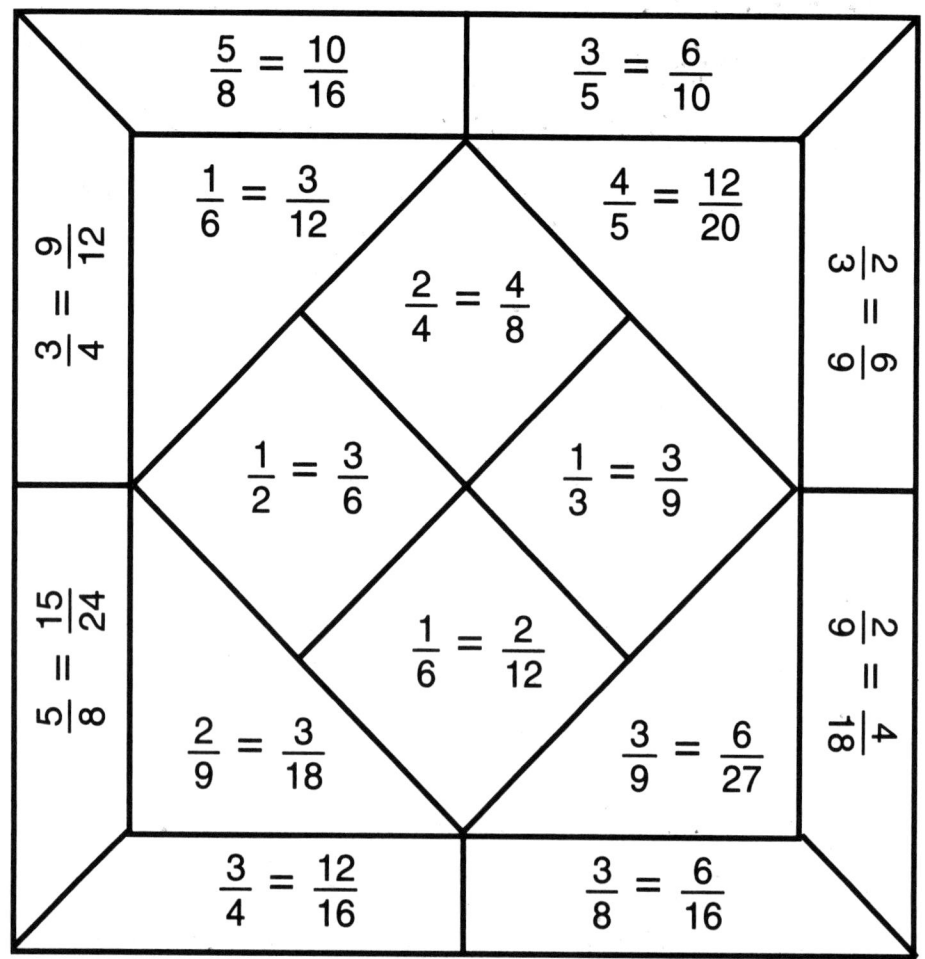

46 Equivalent fractions

FS-23204 Math Made Simple • © Frank Schaffer Publications, Inc.

Bubble Fun

Change each improper fraction to a mixed number. Or, change each mixed number to an improper fraction.

1. $\frac{17}{8}$
2. $1\frac{1}{4}$
3. $\frac{13}{4}$
4. $2\frac{1}{5}$
5. $\frac{47}{9}$
6. $4\frac{2}{5}$
7. $\frac{15}{4}$
8. $3\frac{5}{9}$
9. $4\frac{5}{9}$
10. $\frac{25}{5}$
11. $5\frac{1}{8}$
12. $\frac{12}{3}$
13. $8\frac{7}{9}$
14. $\frac{67}{9}$
15. $\frac{37}{6}$
16. $7\frac{3}{5}$
17. $\frac{83}{5}$
18. $\frac{39}{7}$

Improper fractions and mixed numbers

Let's Compare

Use >, <, or = to compare the fractions below.

A.

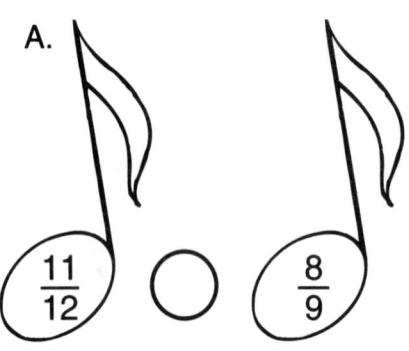
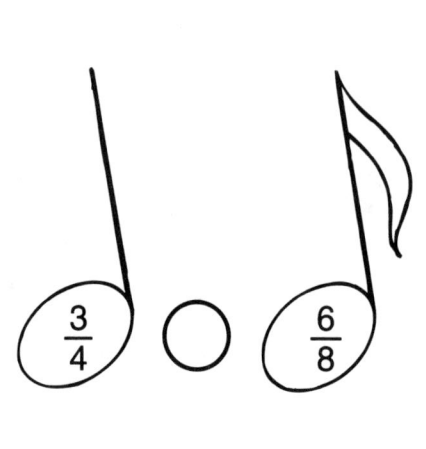
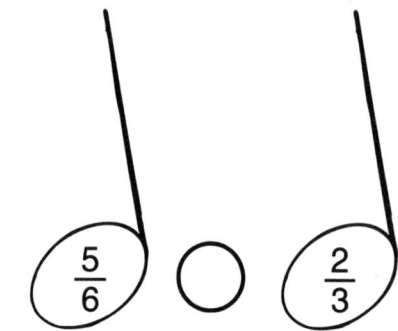

B.

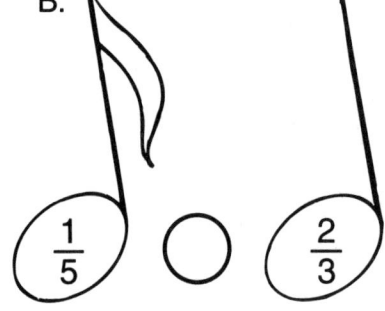
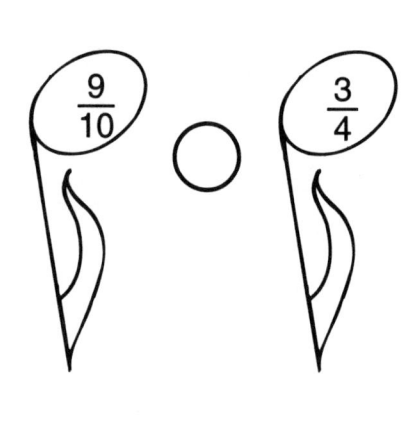
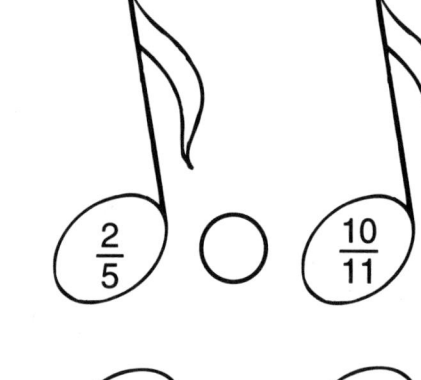

C.

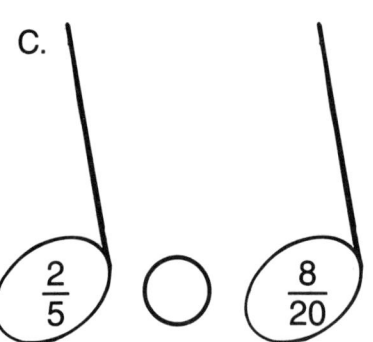
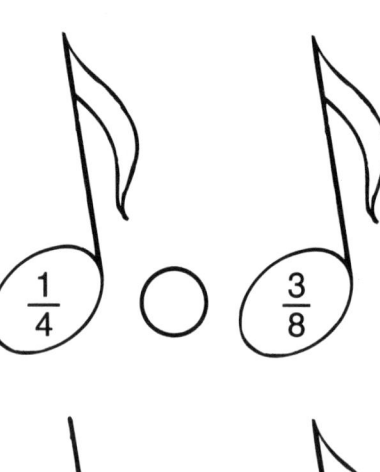
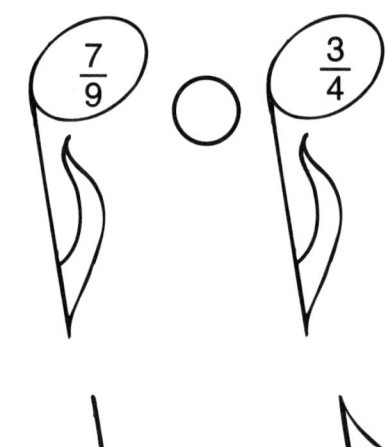

D.

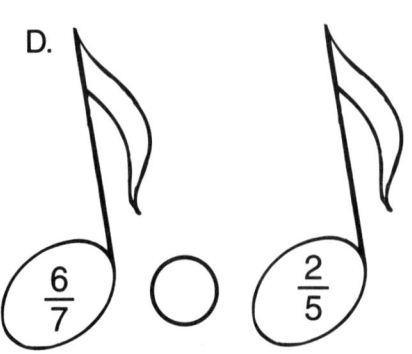
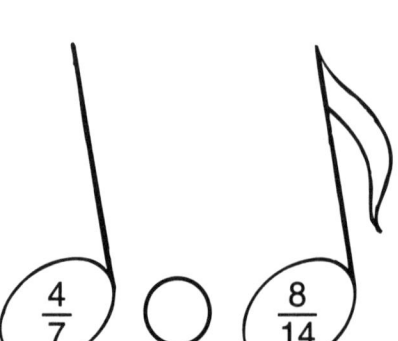
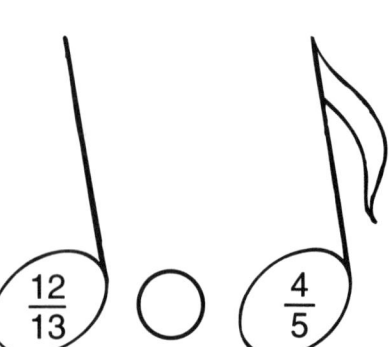

48 Comparing fractions

Cloud Crazy

Add or subtract.

A.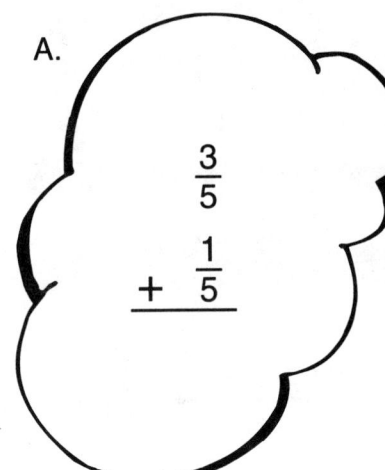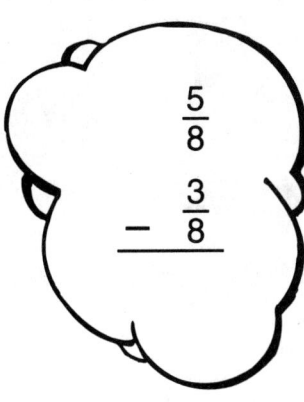

$\frac{3}{5}$ $\frac{1}{8}$ $\frac{1}{4}$ $\frac{5}{8}$
$+\frac{1}{5}$ $+\frac{4}{8}$ $+\frac{2}{4}$ $-\frac{3}{8}$

B.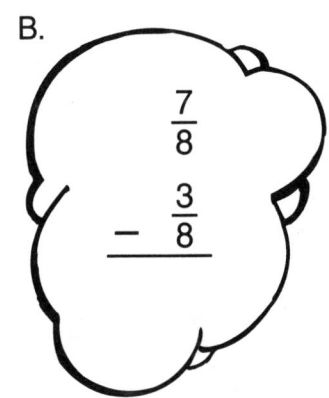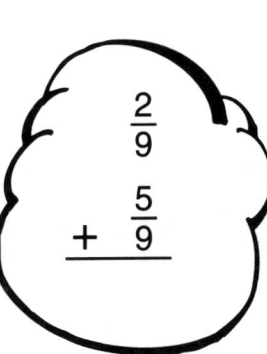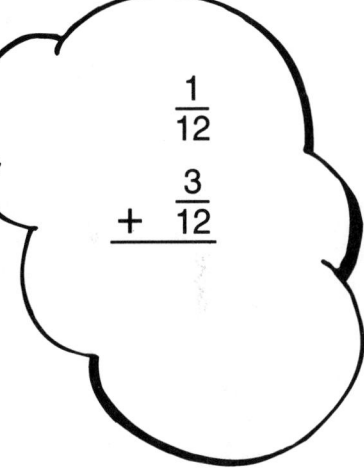

$\frac{7}{8}$ $\frac{2}{9}$ $\frac{3}{4}$ $\frac{1}{12}$
$-\frac{3}{8}$ $+\frac{5}{9}$ $-\frac{1}{4}$ $+\frac{3}{12}$

C. 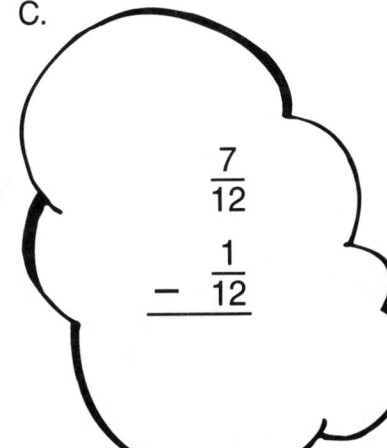 $\frac{3}{8}$ $\frac{5}{9}$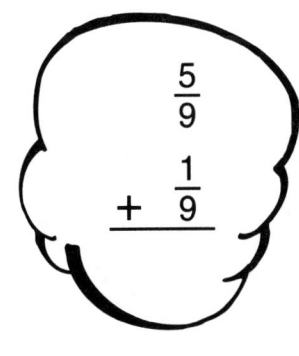

$\frac{7}{12}$ $+\frac{5}{8}$ $+\frac{1}{9}$ $\frac{2}{3}$
$-\frac{1}{12}$ $-\frac{1}{3}$

Adding and subtracting fractions with like denominators

Super Sentences

Complete the number sentences below. Simplify all answers.

$\frac{2}{3}$	+	$\frac{14}{15}$	=	
+	▓	+		
$\frac{1}{30}$	+	$\frac{1}{6}$	=	
=		=		+
				$\frac{1}{40}$
				=

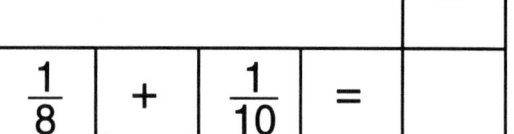

$\frac{1}{8}$	+	$\frac{1}{10}$	=	
+	▓	+		
$\frac{1}{4}$	+	$\frac{1}{5}$	=	
=	▓	=	▓	+
	+		=	
				=

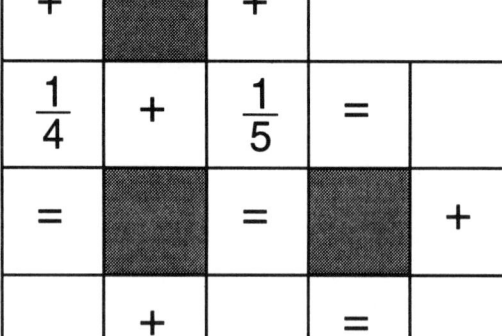

50 Adding fractions with unlike denominators

Fantastic Fruit

Subtract. Simplify all answers.

$\frac{1}{4} - \frac{1}{8}$

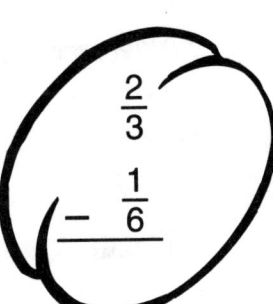
$\frac{2}{3} - \frac{1}{6}$

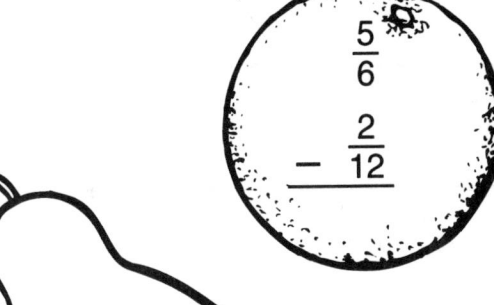
$\frac{5}{6} - \frac{2}{12}$ $\frac{1}{2} - \frac{1}{3}$

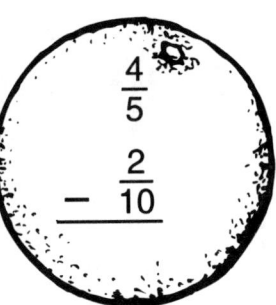
$\frac{4}{5} - \frac{2}{10}$

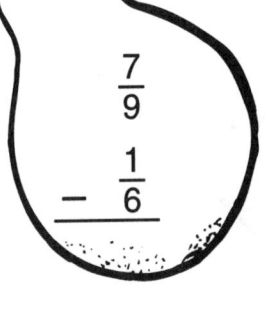
$\frac{7}{9} - \frac{1}{6}$

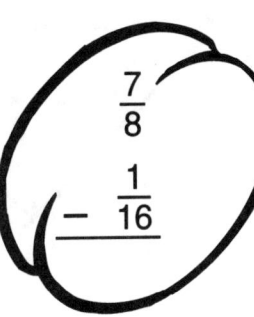
$\frac{7}{8} - \frac{1}{16}$

$\frac{3}{4} - \frac{1}{12}$

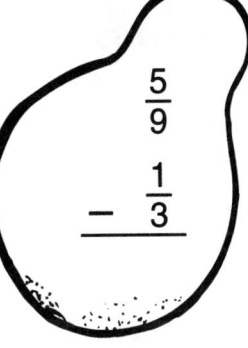
$\frac{5}{9} - \frac{1}{3}$

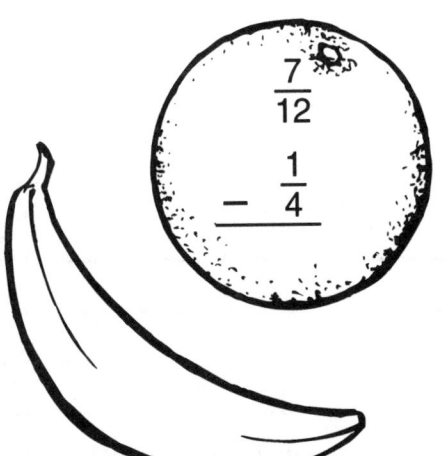
$\frac{7}{12} - \frac{1}{4}$

$\frac{11}{21} - \frac{1}{6}$

$\frac{7}{12} - \frac{3}{18}$

Subtracting fractions with unlike denominators

Geometry

With the help of the activities in this section, students will manipulate models of solids to develop an understanding of their properties and spatial relationships. They will also begin to develop such important skills as perimeter, area, and volume. These experiences will give students a background for more advanced geometry presented in future years.

Be sure to point out how students may use their new skills in everyday life. For example, students will use their knowledge of geometry when they construct geometric figures or designs, or use symmetry in art. Help students observe the world around them and identify their own connections to geometry.

CONCEPTS

The ideas and activities presented in this section will help students explore the following concepts:
- plane figures
- congruent figures
- circles
- perimeter
- volume
- space figures
- symmetry
- lines and angles
- area

TANGRAMS — Manipulative Activity

Read or tell the story of Grandfather Tan and the tangrams. Fourth-graders can make the animals in the book while you read. Using cardstock, make copies of the tangram pattern to the right for students to use. After they have made the figures in the story, students can design their own figures using all seven pieces. They can trace around the outside of their shapes on pieces of paper. Have students exchange shapes and complete each other's. Some students can pair up with friends and make pictures using double sets.

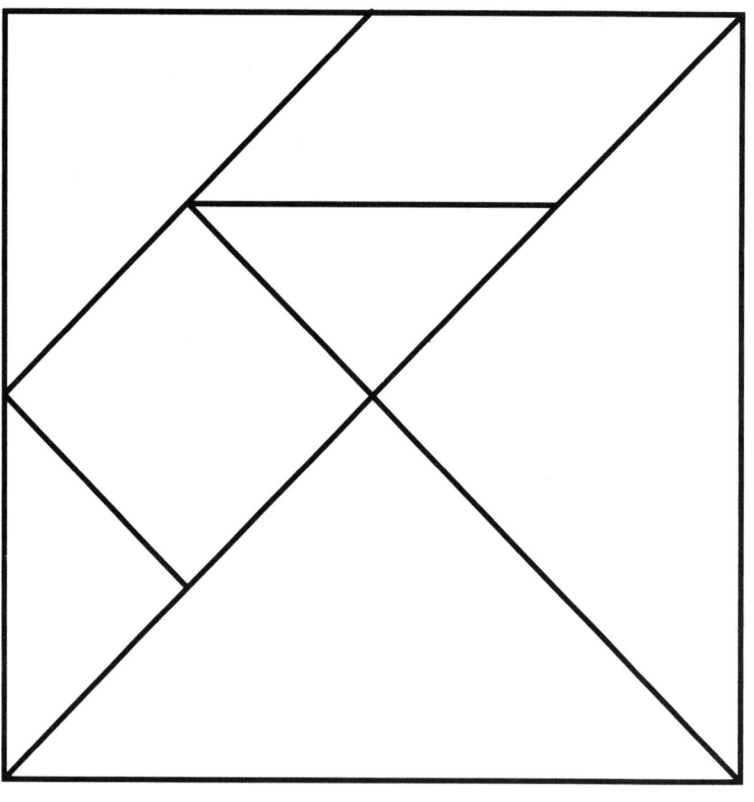

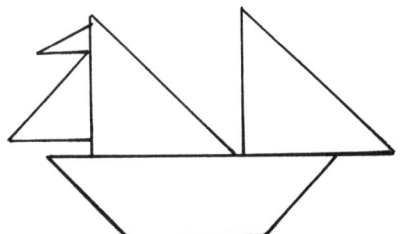

LINES — Writing Activity

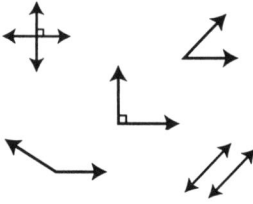

Students can draw and label lines (intersecting, parallel, and perpendicular), line segments, rays, and angles. They can also make the drawings in their journals, for future reference, and can write the definitions beside the drawings.

RIGHT ANGLES *Writing Activity*

Right angles can be found around the room by using the corner of a sheet of paper or cardstock. Students can look for right angles and list the places they find them (tables, windows, index cards, books, etc.). In their journals, have students write a definition for *right angle* and list at least three examples of where one can be found.

PERIMETER *Class Activity*

To determine perimeter, have students use meter sticks, rulers, or strips of yarn that can be measured. They can make a list of items they measure and their perimeters in their journals.

When they are somewhat familiar with lengths, students can estimate perimeters and then actually measure to find out how close their estimates are.

BASIC SHAPES *Writing Activity*

Review basic shapes with students. Students can draw and describe *circles, triangles, squares, rectangles, parallelograms, pentagons, hexagons,* and *octagons* in their journals. Other new words to add include *face, edge,* and *vertex*.

Have the students bring objects made from the shapes or pictures showing objects made from the shapes. Display them on a bulletin board. Simple items such as an ice cream cone, a soup can, an orange, and cereal or other boxes will give you a good start. Have students match the shapes with the names written on index cards.

USING ATTRIBUTE BLOCKS *Manipulative Activity*

Discuss the meaning of *attribute* with the students and then brainstorm with them the different attributes of the blocks (thickness, color, shape, size). Begin with one shape, telling about its attributes. Have a student who has a similar block with only one different attribute describe his or her block. Continue until everyone has shared his or her block. On another day, ask for two differences, three differences, etc.

Congruent Shapes *Class Activities*

Have students guess which shapes are congruent as you show pictures or display cutout figures in different directions. They can then lay one on the other to check their guesses.

Have the students make a list of things that are congruent such as stars in a box, stickers, stamps, etc. In their journals, students can use cutout shapes, and trace around them to be sure the drawings are actually congruent.

SYMMETRY — Class Activities

- Have students fold sheets of paper in half and cut through both thicknesses of paper to create a variety of symmetrical shapes that you can display around the room.
- In nature, there are many examples of symmetry. Leaves and stems with leaves are the easiest examples to find. Butterflies and flowers also often demonstrate the idea.
- Have each student draw half of a figure, exchange his or her paper with a friend, and have the friend finish the design symmetrically.
- Let each student use pattern blocks to create half of a picture or a shape, and have a classmate finish the other side in reverse, so that it is symmetrical. Each student can then trace his or her half on paper and give it to a classmate or put it in a center for another student to complete.
- Using geoboards to design halves of shapes is fun for students. They can then trade boards and have other students design the symmetrical other halves.

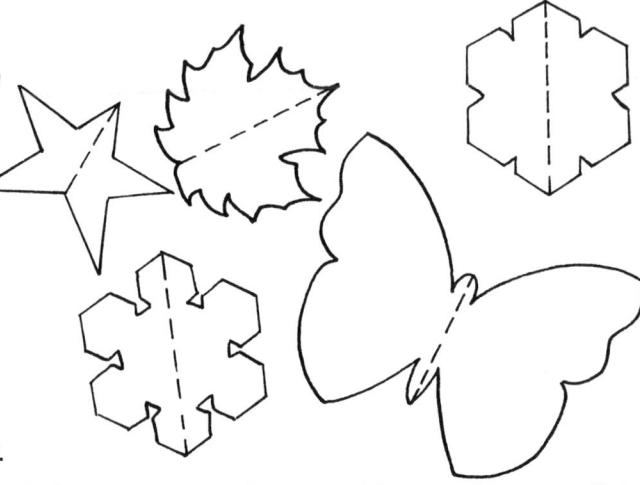

AREA — Manipulative Activities

Students can begin to find the area of things around them. As they use graph paper and count units, students may discover, or you can lead them to the idea, that there is a quick way to find area using multiplication. Give them practice measuring lengths times widths, and have them write this formula in their journals: *Area = length x width.*

Give students graph paper to color in various areas. Students can also use the graph paper to experiment finding different ways to show the same area with different perimeters.

VOLUME — Class Activities

Actual building with linking cubes gives students the best possible view to help them understand pictures that show volume. Without this understanding, students often only count the faces that show.

You will probably need to help students see the extension of the area formula into the volume formula (length x width x height) to get the volume of an object.

Have students predict the number of cubes that will fit into a box or other container and then actually fill the container, placing the cubes close together, to find the volume. They can try measuring the container to help with their predictions.

Homework — School-Home Connection

Have students classify the objects in their refrigerators into groups of solid figures such as cylinders, cubes, rectangular prisms, and combination figures. Then students may use estimation to determine which type of solid figure is most used for food packaging (i.e., liquid—cylinders; solids—rectangular prisms).

Name _____

Matching

In Shape

Match the names with the shapes.

Plane Figures

1. △ 2. ▢

3. ⬡ 4. ⬠

5. ▭ 6. ⬗

7. ▱ 8. ▱

9. △

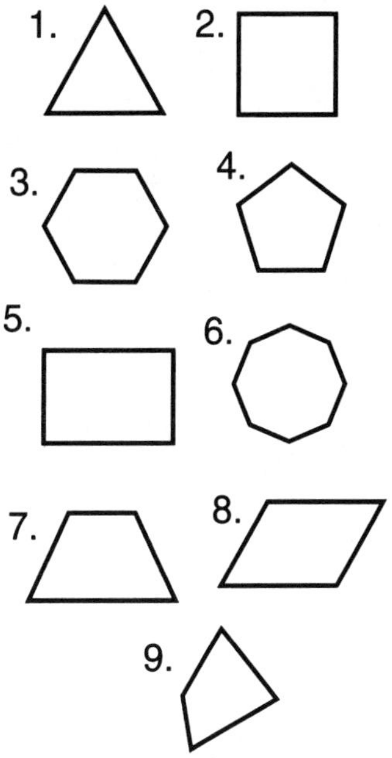

____ hexagon
____ square
____ rectangle
____ octagon
____ parallelogram
____ triangle
____ pentagon
____ trapezoid
____ quadrilateral

Space Figures

10. 11. 12. 13. 14. 15.

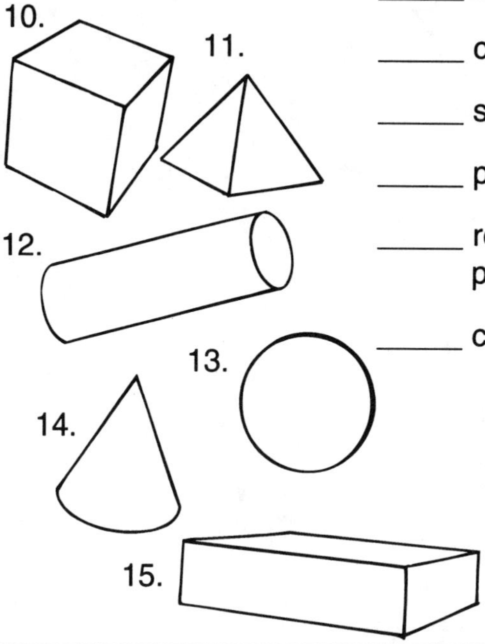

____ cone
____ cylinder
____ sphere
____ pyramid
____ rectangular prism
____ cube

16. On a separate sheet of paper, draw a picture that contains at least one of every figure. Label the figures.

Each flat surface is a *face*. Two faces meet at an *edge*. Three (or more) edges meet at a *vertex*. Label the parts shown.

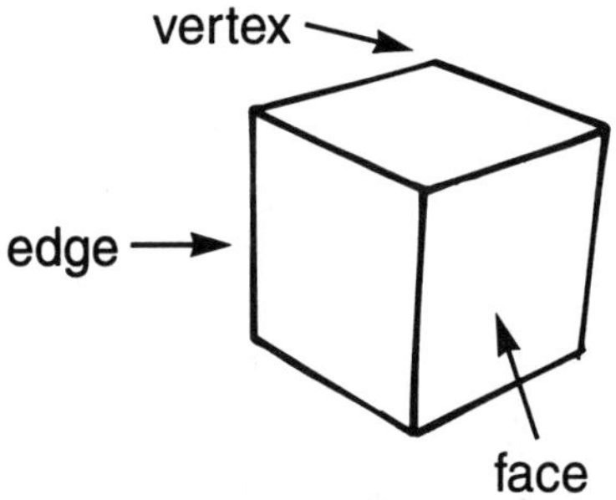

vertex →
edge →
face

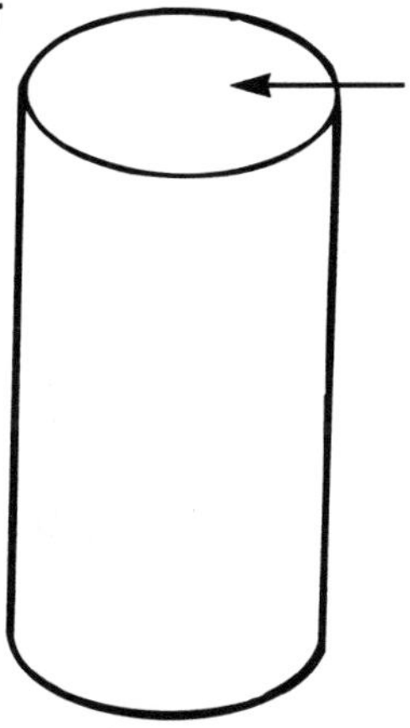

17. 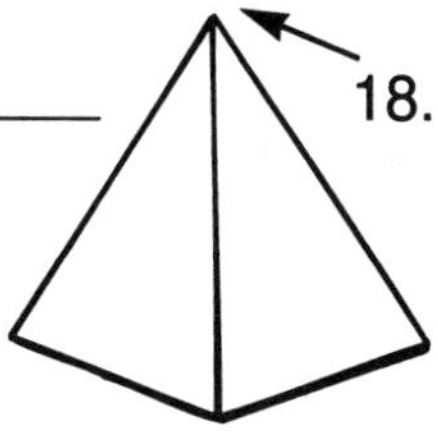 _____   18. _____

19. _____

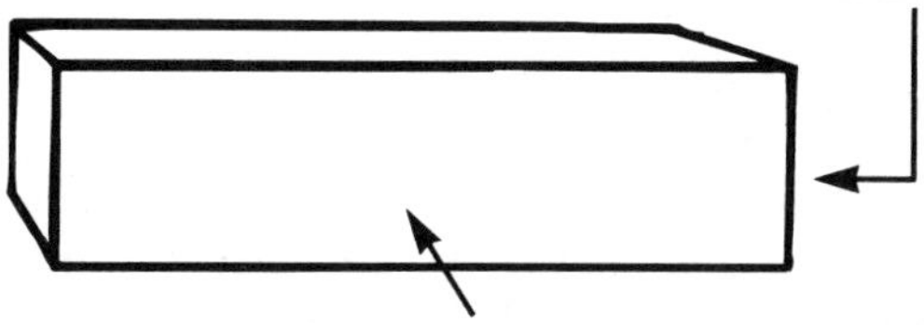

20. _____

Write the number of sides and vertices for each figure.

21.  △ ____ sides ____ vertices

22. 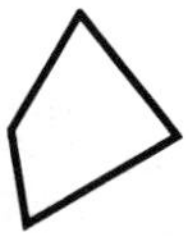 △ ____ sides ____ vertices

23.  ⬡ ____ sides ____ vertices

Geometric shapes

Name_____

It's the Same

A. Congruent figures have the same size and shape. Color the congruent figures.

B. A line of symmetry divides a shape into congruent shapes. Color only the figures with lines of symmetry.

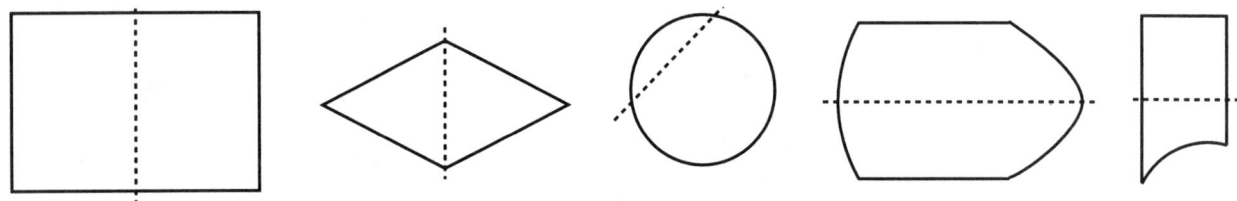

C. Use a ruler to draw two sets of congruent figures. Then draw two figures, each with a line of symmetry.

Congruent Figures	Symmetry

D.

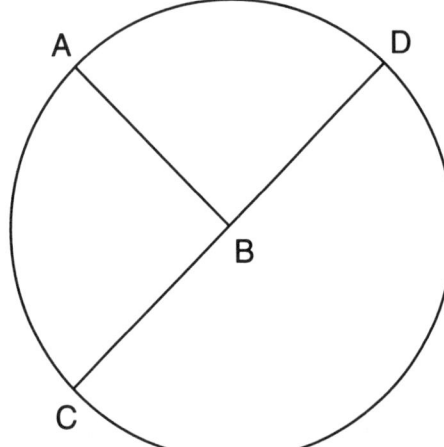

1. A circle is named by its center.

    This circle is _____.

2. Name 2 radii. _____

3. Name a diameter. _____

4. Draw another diameter using a red pencil or crayon. Label it EF.

5. Draw two radii using a blue crayon. Label them BG and BH.

56 Congruency; symmetry; circles   FS-23204 Math Made Simple • © Frank Schaffer Publications, Inc.

Angles and Lines

Acute angles measure less than 90°.

Right angles measure 90°.

Obtuse angles measure more than 90° but less than 180°.

Parallel lines never intersect.

Perpendicular lines intersect at 90° angles.

Use the figure below to answer the questions.

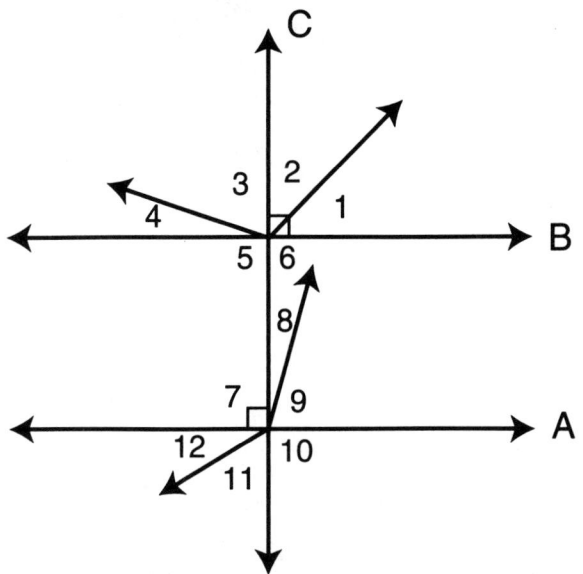

1. Name a pair of parallel lines. _____
2. Name four acute angles. _____
3. Name two right angles. _____
4. Name a pair of perpendicular lines. _____
5. Angle 11 is a(n) _____ angle.
6. Angle 7 plus angle 8 make a(n) _____ angle.
7. Line A is perpendicular to line _____.
8. Angle 1 plus angle 2 make a(n) _____ angle.
9. Draw an obtuse angle.   10. Draw a right angle.   11. Draw an acute angle.

Perimeter Problems

What is the highest waterfall in the world?

To find out, find the perimeter of each figure below at the bottom of the page. Write the letter that represents each problem above its answer.

Remember: Perimeter is equal to the sum of the sides of a figure.

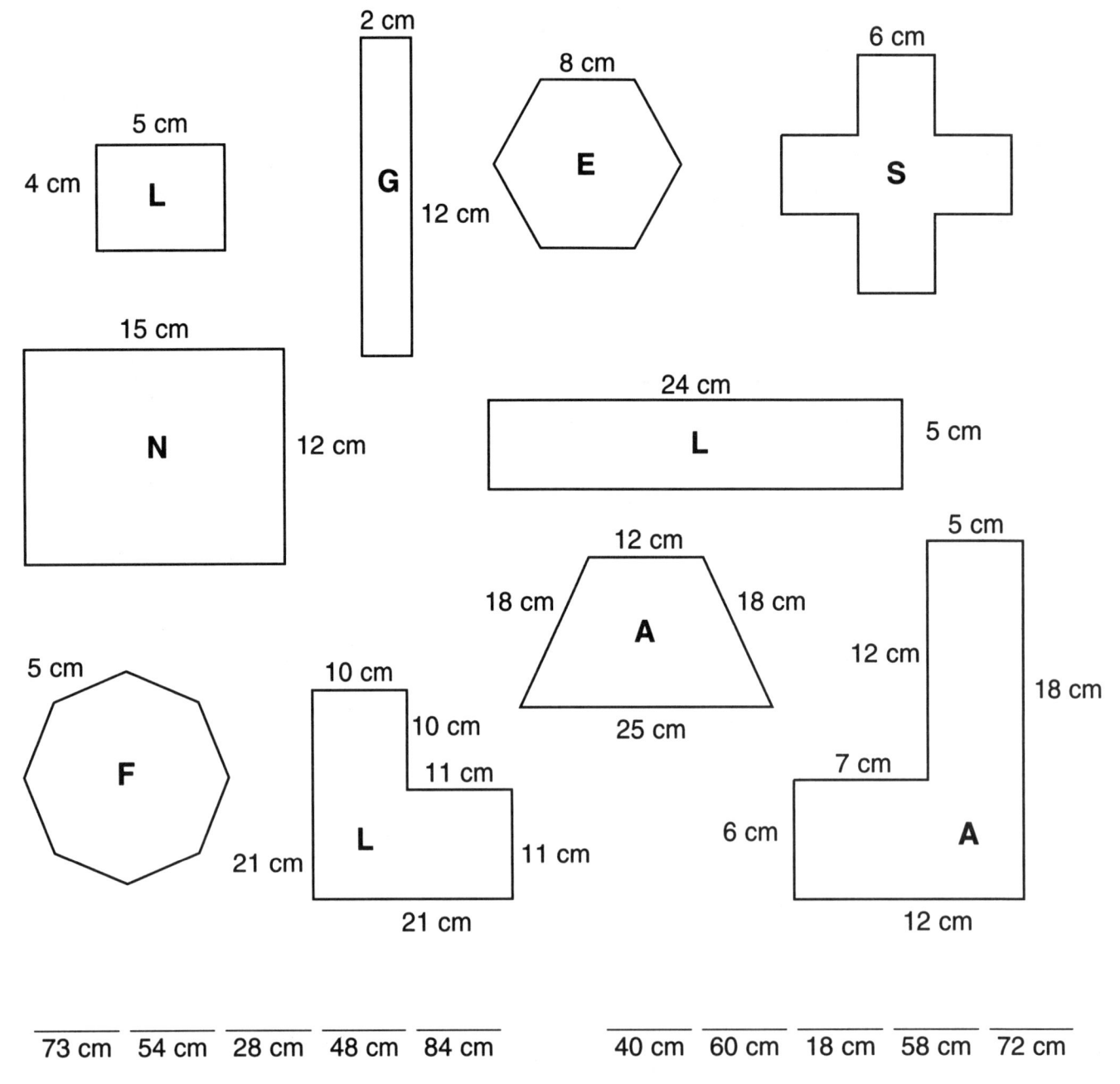

__73 cm__ __54 cm__ __28 cm__ __48 cm__ __84 cm__   __40 cm__ __60 cm__ __18 cm__ __58 cm__ __72 cm__

Analyzing Area

Area tells the number of square units in a figure. Find the area of each figure below.

1. 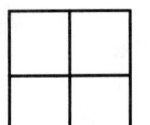 _____ sq. units

2. 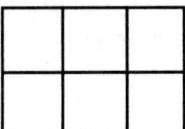 _____ sq. units

3. 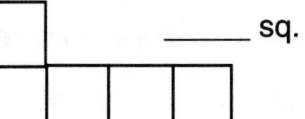 _____ sq. units

4. 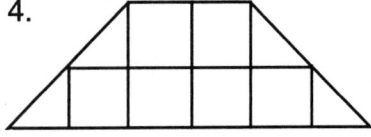 _____ sq. units

5. 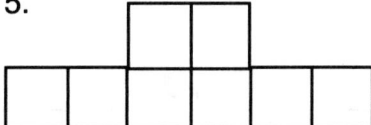 _____ sq. units

The area can be found by multiplying the length times the width. A = lw

6. 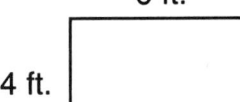 _____ x _____ = _____ ft.^{2}

7. 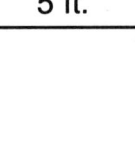 _____ x _____ = _____ ft.^{2}

Find the areas below.

8.  A = _____

9. 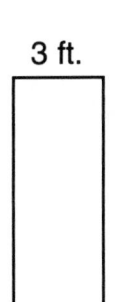 A = _____

10.  A = _____

11.  A = _____

12.  A = _____

13.  A = _____

14.  A = _____

Amazing Muscles

It takes 17 muscles to smile. How many muscles does it take to frown?

To find out, find the areas at the bottom of the page.
Write the letter that represents each problem above its answer.

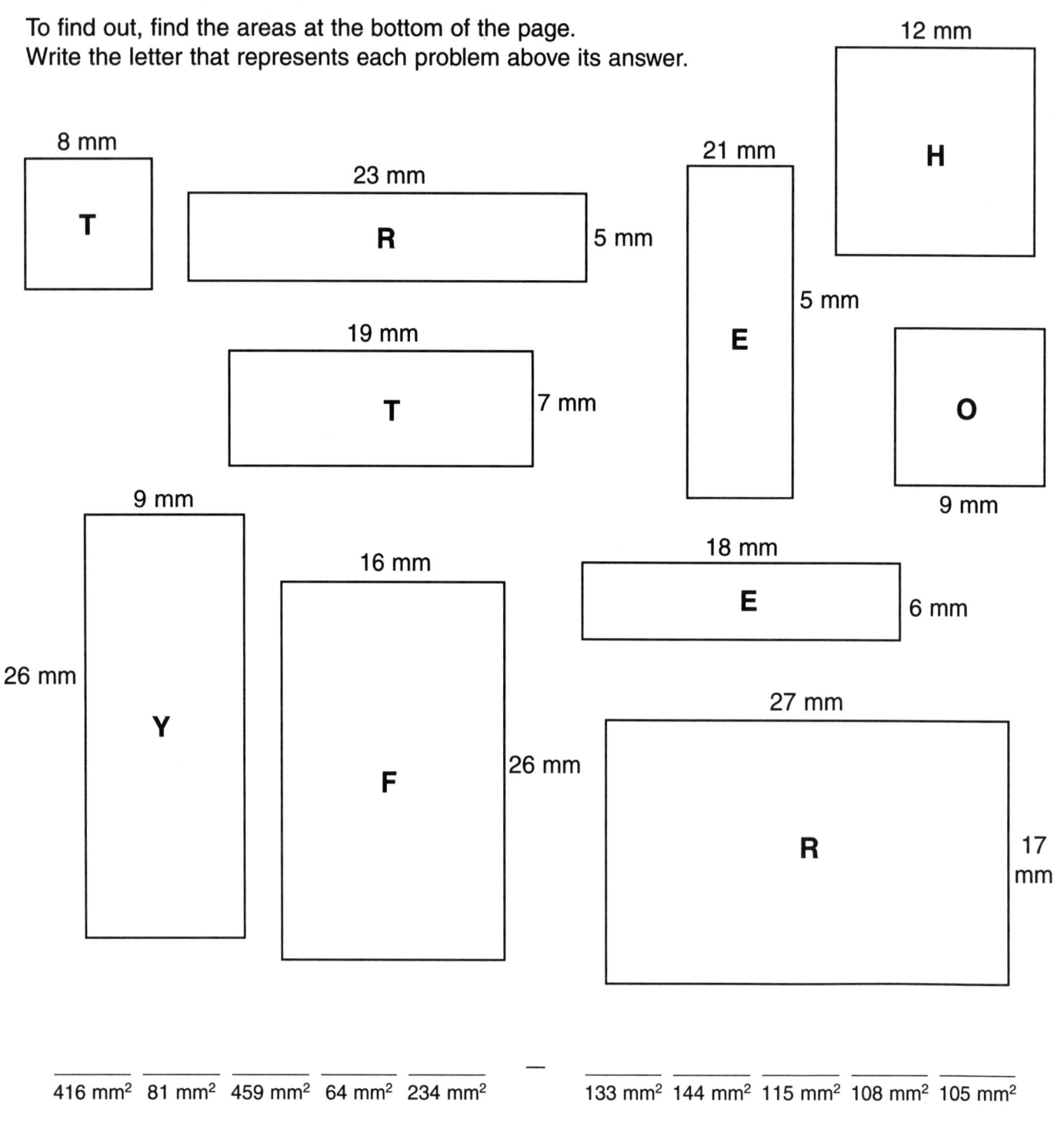

___ ___ ___ ___ ___   ___ ___ ___ ___ ___
416 mm² 81 mm² 459 mm² 64 mm² 234 mm²   133 mm² 144 mm² 115 mm² 108 mm² 105 mm²

Name_____

Very Important Volume

Volume is the number of cubic units that will fit inside a space figure. Build each figure below using linking cubes. Tell the number of cubes used.

1.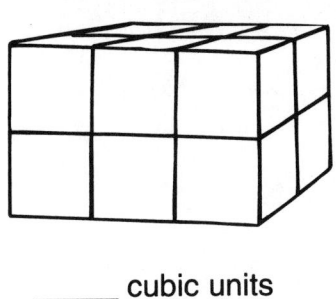

    _____ cubic units

2.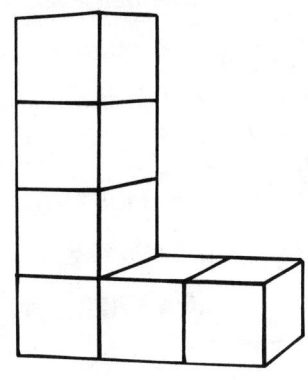

    _____ cubic units

3.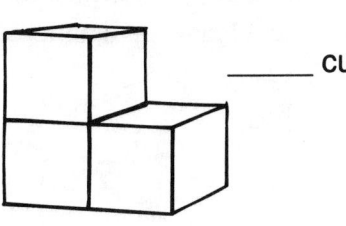

    _____ cubic units

Volume can be found by multiplying the length times the width times the height.

$$V = lwh$$

Find the volume of each figure below.

4.

    _____ x _____ x _____ = _____ cu. units

5.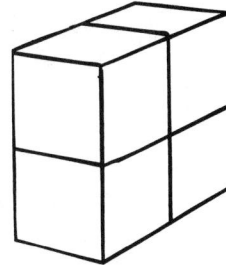

    _____ x _____ x _____ = _____ cu. units

6.

    _____ x _____ x _____ = _____ cu. units

7.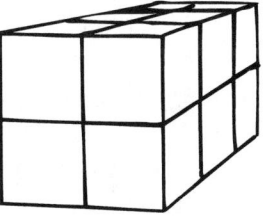

    _____ x _____ x _____ = _____ cu. units

8.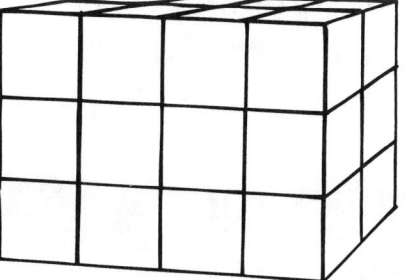

    _____ x _____ x _____ = _____ cu. units

9.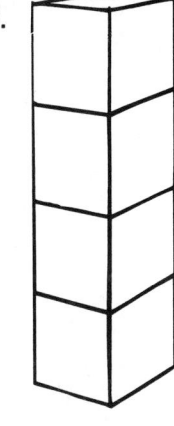

    _____ x _____ x _____ = _____ cu. units

FS-23204 Math Made Simple • © Frank Schaffer Publications, Inc.

Measurement

Measurements are used in daily life to help people understand, organize, and compare objects and materials found in their environments. Students are provided with a wonderful variety of activities involving measurement below and on pages 63–66.

CONCEPTS

The ideas and activities presented in this section will help students explore the following concepts:
- metric measurement
- standard measurement
- linear measurement
- weight measurement
- liquid measurement
- temperature

LINEAR MEASUREMENT — Class Activities

Students love to measure things, so they'll have fun learning during measurement experiences. They can use metric rulers and standard rulers. (It is a good idea to have both types of rulers available for comparisons.) You can make a lesson out of questions such as "Approximately how many centimeters are in 20 inches?" Students can compare using the rulers and begin to get general ideas about converting measurements.

Give students a lot of opportunities to measure things at their desks (books, pencils, erasers, etc.), their desks themselves, and other things around the room. Set up a center where students can estimate lengths of objects and then measure them.

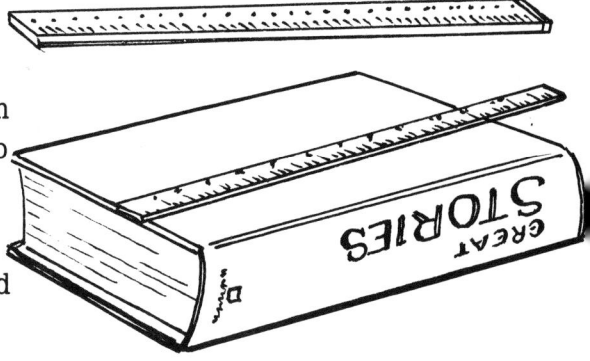

EQUIVALENT AMOUNTS — Writing Activity

Make a chart and have students copy it into their journals for reference. Include the following: 1 m = 100 cm, 1000 mm = 1 m, 10 mm = 1 cm, etc. Do the same for these standard measurements: 12 in. = 1 foot, 3 feet = 1 yard, etc. In spare minutes, ask questions such as "How many feet are in 6 yards?", "How many centimeters are in 3 meters?", etc.

MEASUREMENT ART — Art Activity

On a chart, draw line segments such as 1 cm in red, 2 cm in blue, 3 cm in green, 4 cm in purple, etc. Then have the students create pictures using the line segments. Students need to decide on a segment's length and actually draw it in the color you have shown on your chart.

When done, students can multiply each length by the number of segments and add to find the total number of centimeters used in their pictures.

WEIGHING THINGS — Class Activity

If possible, have students weigh things to get a good concept of weights. You can tell them that grams are used for light objects like paper clips or coins, and that heavier objects, such as books, are weighed in kilograms. Students can draw pictures of light objects and heavy objects in their journals.

Encourage parents to take their children to the store and have them help weigh fruits and vegetables. They can also read and compare weights on packages. You can also set up a center in your room with containers that show different weights. For example, students could share the weights of their snacks on a given day.

TEMPERATURE — Class Activities

- Keep a thermometer hanging in your room. As the weather changes, have the students estimate the temperature. You can graph the temperature for a few days each month to compare as the year goes on.

- Students may be able to compare degrees Fahrenheit and Celsius. Some thermometers show both measurements. Students should begin to realize that ⁻5°F is very cold. Students who live in warm states may never have been exposed to weather extremes. Have students make drawings of different activities and list possible temperatures to give students a better sense of temperatures.

LIQUID MEASUREMENT — Writing Activity

Set up a row of containers of various sizes such as liter and half-liter bottles. Have students compare the different containers. Students can write the comparisons in their journals. Students will enjoy pouring cups of water into pint, quart, and gallon containers from the liter containers. This will really help them see the relationships of the different amounts.

MEASURING WITH UNITS — Manipulative Activity

Have students use units such as hand spans, paper clips, etc., to measure things. Have the students estimate how many paper clips will be needed to measure a book and then actually measure it. Estimates should improve as they continue to measure objects.

Name _____

Let's Measure

millimeters, centimeters, kilometers	grams, kilograms	milliliters, liters
10 mm = 1 cm 100 cm = 1 m 1,000 m = 1 km	1,000 g = 1 kg	1,000 mL = 1 L

Measuring Length and Distance

Very small things are measured in millimeters (mm). A mm is $\frac{1}{10}$ of a centimeter (cm). Write the names of 3 small things. Measure them in mm. Then list and measure 3 things in cm.

millimeters   1. _____   2. _____   3. _____

centimeters   1. _____   2. _____   3. _____

Large things and distances are measured in kilometers (km). Next to each item, write **mm**, **cm**, or **km** to show what you would use to measure it.

A. small leaf _____   E. width of a large lake _____

B. crayon _____   F. small paper clip _____

C. spool of thread _____   G. your math book _____

D. distance to New York _____   H. a corn kernel _____

Weighing Things

Grams (g) are used to weigh light objects such as paper clips or coins. Kilograms (kg) are used to weigh heavier objects. Choose **g** or **kg** for each object.

I. a postcard _____   J. your dog _____   K. a van _____   L. a mouse _____

Liquid Measurement

Small amounts of liquids are measured in milliliters (mL). Choose **mL** or **L** for each amount below.

M. a small cup of soup _____

N. medicine in a spoon _____

O. water in a large pail _____

P. water in a swimming pool _____

Q. milk in a glass _____

R. gas in a car _____

Name_____

How Much Is There?

inches, feet, yards, miles	ounces, pounds, tons	cups, pints, quarts, gallons
12 in. = 1 ft.	16 oz. = 1 lb.	2 c. = 1 pt.
36 in. or 3 ft. = 1 yd.	2,000 lb. = 1 T	2 p. = 1 qt.
5,280 ft. = 1 mi.		4 qt. = 1 gal.
1,760 yd. = 1 mi.		

Measuring Length and Distance

Write **inches**, **feet**, **yards**, or **miles** to show how you would measure:

1. a crayon _____
2. a pencil _____
3. distance to another city _____
4. a piece of cloth for a dress _____
5. length of the room _____
6. person's height _____

Weighing Things

Write **ounces**, **pounds**, or **tons** to show how you would measure:

7. a block of cheese _____ or _____
8. a peach _____
9. some pepper _____
10. a sack of potatoes _____
11. a stick of butter _____
12. a truck _____

Liquid Measurement

Write **cup**, **pint**, **quart**, or **gallon** to show how you would measure:

13. a glass of water _____
14. a swimming pool _____
15. a bathtub _____
16. a large jug of milk _____
17. a bucket of paint _____
18. whipping cream _____

Write the units you would use for each measurement in the boxes below. Use the metric and standard units of measurement.

Length		Weight		Liquid	
Metric	Standard	Metric	Standard	Metric	Standard
1. ____	1. ____	1. ____	1. ____	1. ____	1. ____
2. ____	2. ____	2. ____	2. ____	2. ____	2. ____
3. ____	3. ____		3. ____		3. ____
	4. ____				4. ____

Standard measurement

Name_____

How Hot Is It?

Temperature is measured in degrees Celsius or Fahrenheit.

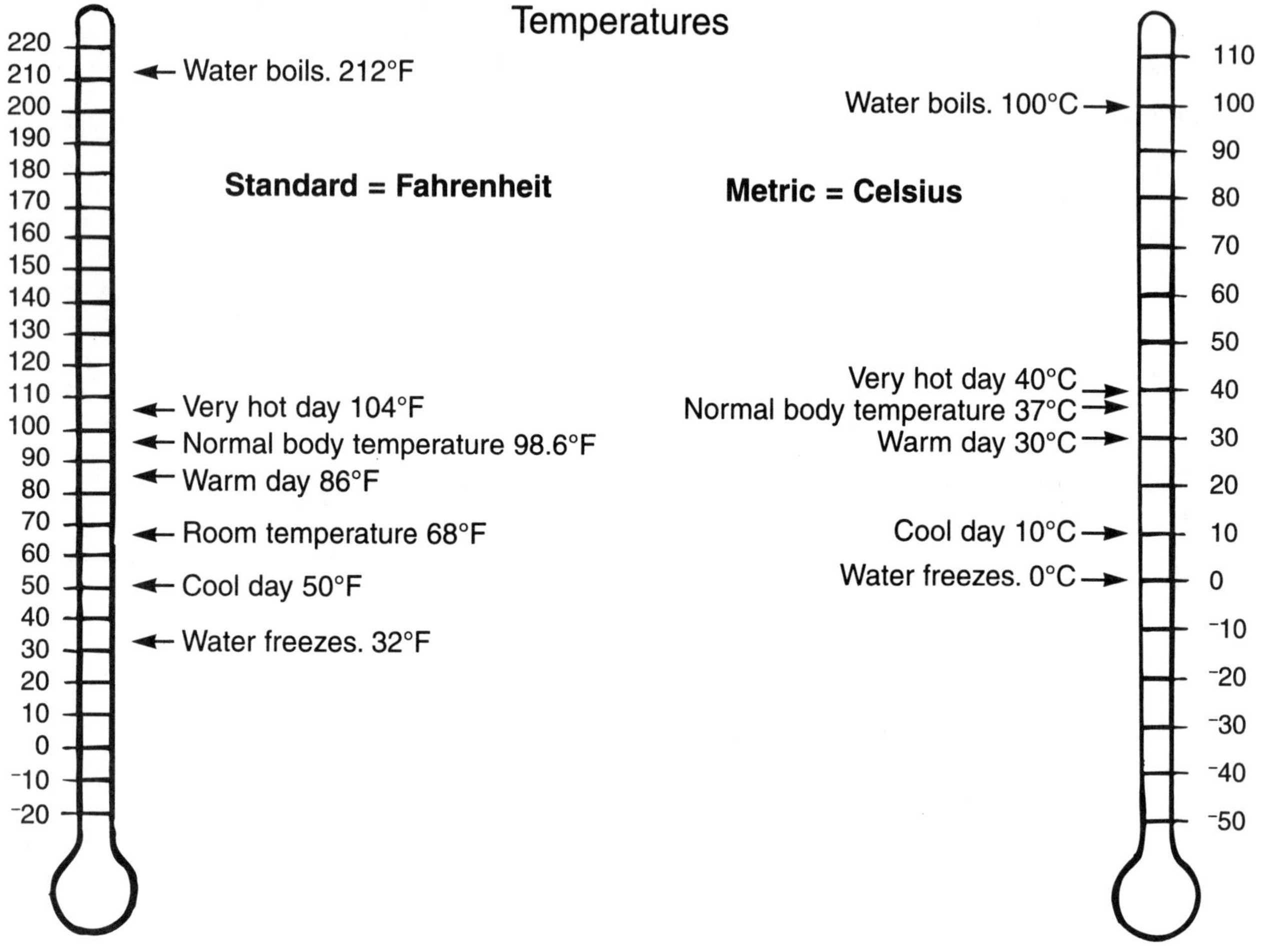

Temperatures

Standard = Fahrenheit

- Water boils. 212°F
- Very hot day 104°F
- Normal body temperature 98.6°F
- Warm day 86°F
- Room temperature 68°F
- Cool day 50°F
- Water freezes. 32°F

Metric = Celsius

- Water boils. 100°C
- Very hot day 40°C
- Normal body temperature 37°C
- Warm day 30°C
- Cool day 10°C
- Water freezes. 0°C

Use the thermometers to show the temperatures below in Celsius and Fahrenheit.

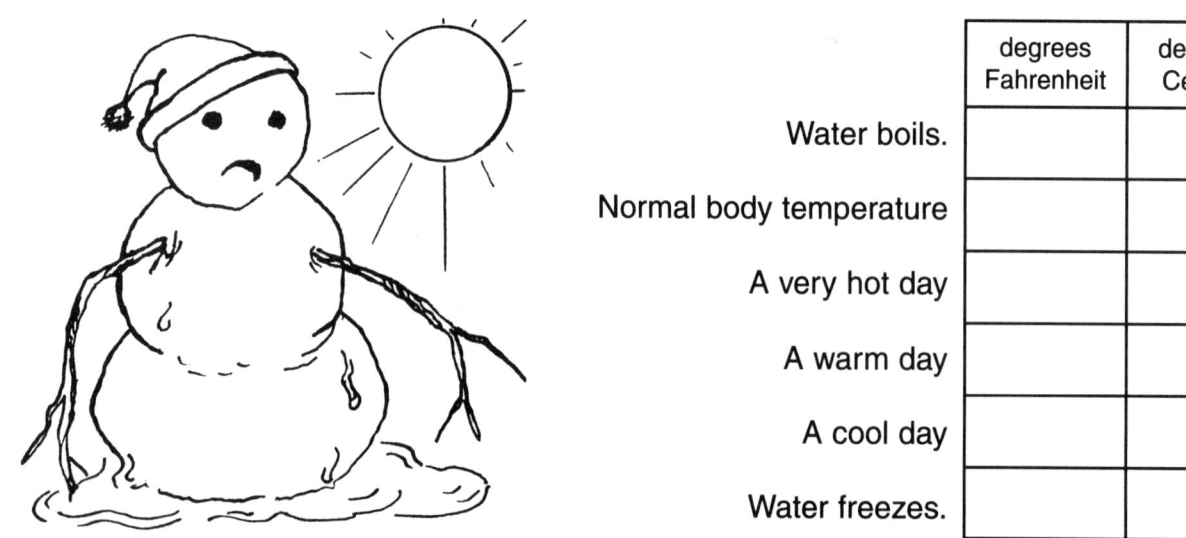

	degrees Fahrenheit	degrees Celsius
Water boils.		
Normal body temperature		
A very hot day		
A warm day		
A cool day		
Water freezes.		

66 Temperature—Celsius, Fahrenheit

FS-23204 Math Made Simple • © Frank Schaffer Publications, Inc.

Probability and Graphing

Included in this section are activities that allow students to collect and organize data, and to help them see concrete relationships between math and real life. Skills related to creating and interpreting data in pictographs, bar graphs, tables, and charts can be used in many subject areas, as well as in daily activities. Classroom and individual experiments will help students predict probability outcomes.

CONCEPTS

The ideas and activities presented in this section will help students explore the following concepts:
- statistics and probability
- charts
- line graphs
- bar graphs
- circle graphs
- coordinate graphs

CLASS AND SMALL GROUP GRAPHS — *Group Activity*

Have students brainstorm and list the things they can graph such as favorite things (colors, sports, authors, etc.). Divide students into small groups. Have each group decide on a topic to collect information on and use the information to make a graph. They may interview friends and family members for opinions and choices, or find some information in the newspapers or on the Internet. After gathering information, students may need to tally the results before transferring it to their graphs. You should remind them that they can have a choice entitled "other" to include opinions that were expressed by just a few people, or even by just one person.

Next, students should decide on the best type of graph to use to show their information. A simple bar graph, horizontal or vertical, may be good. For a presentation to the class or other classes, a pictograph may be more visually interesting.

LITERATURE CONNECTION — *Manipulative Activity*

Have each student locate a paragraph in a favorite story and tally the number of 1-, 2-, 3-, 4-, 5-, and 6-letter or more words there are in the paragraph. They can then graph the results and compare them with someone who studied the same paragraph (the results should be the same). Or, students can compare their graphs with someone who tallied a different paragraph. Students can also begin to discuss why some books have long words and some are limited to short words (grade level difficulty, controlled vocabulary, etc.).

FS-23204 Math Made Simple ■ © Frank Schaffer Publications, Inc.

PROBABILITY GAMES

Games

Playing probability games will be one of the favorite activities in your class.

- Have the students make spinners. One spinner should have just one each of four different items making the probability 1 in 4. The other spinners should have different amounts of items (letters, colors, etc.) so that the probability will change. Therefore, if there are two blue areas out of four, the probability will change to 2 out of 4.

  Have students predict the number of times each item on the spinner (color, letter, etc.) will appear out of 10. Then have the students spin 10 times, record the results, and compare these with their predictions. Have the students repeat the experiment several times to see how close they came to their predictions. Also explain or discuss at this time why scientists never rely on only one experiment for results.

- Repeat this type of experiment using different colors or objects placed in a bag. Students can predict how many of each object will be drawn by members of the group. After each person has drawn and replaced the object in the bag, students should tally and chart the results and then discuss differences. They will note that although they will probably have the results expected over a period of several tries averaged out, they may not get the expected results every time. This is a good subject of reflection for students' journals.

- Place all of the students' names in a bag and write the number of boys and the number of girls on the board. Write the ratio of boys to the whole class, and the ratio of girls to the whole class. Have students predict the outcome if everyone draws and replaces the names. Students should come fairly close to determine how many times boys' or girls' names will be drawn.

- Put the days of the week in a bag. Ask students for the odds of picking any one day (1 in 7), a weekend day (2 in 7), and a weekday (5 in 7). Have students figure odds on days beginning with S or T as well as those beginning with M, W, or F.

CHARTS

Class Activity

Information is often recorded on charts. Obtain bus, train, or plane time schedules from local companies. Write questions for students to answer using the schedules (for example, "What is the earliest time you can catch a bus headed for Las Vegas on Saturday? If you miss that bus, how long will you have to wait for the next one?").

The daily schedule for your class can stimulate students to write questions for others to answer (for example, "How many minutes are there between recess and lunch? How many minutes before math did we begin reading?", etc.).

ESTIMATING CONTAINERS *Manipulative Activity*

One activity that students will enjoy is estimating the number of objects in a jar or other container. Students can predict how many of an object, such as sunflower seeds, are in a container. They can find the average prediction of the class and then count to see how close they came. If they aren't anywhere near the total, have them brainstorm reasons why and then see if their next predictions are closer.

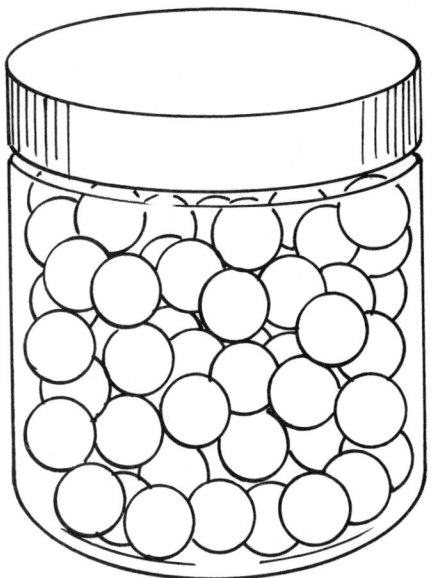

Each time you fill the jar with new objects, students should base their new predictions of the amount of objects in the jar on the things they know, such as the size and shape of the new object compared to the old object (sunflower seeds). As the year goes on, change the amounts in the jar, by perhaps filling it halfway. Students should discuss the reasons for the predictions and then see if they are getting closer to the actual totals.

You can send a letter to parents explaining this activity and asking for parent volunteers to fill the jar with objects at different times. Be sure to ask if they want the contents returned after the items have been counted. If they are placing very small items, such as rice, in the jar, tell them that they might want to fill the jar only ¼ of the way full. If the counting gets tiresome, just divide it into a small amount per student and have students add their amounts to the running total.

You can have individual predictions listed and have a winner of the month whose prediction was closest. Keep a chart of winners. If the same person wins more than once, have a second-place winner that month to include as many students as possible in winning.

VOCABULARY *Writing Activity*

Have students group words in their math journals, including words such as *sure*, *definite*, and *certain* if the odds are even, or *probably*, when odds are close. They can go for *uncertain*, *maybe*, or *possibility*, when odds are unclear.

School-Home Connection *Homework*

Have the students and their parents search through newspapers and magazines for charts and graphs. News magazines and business sections of papers and magazines often have visual presentations of information. Set up a bulletin board with the graphs and charts brought to class. You and the students can write questions about the charts or graphs, and use them as group or center activities. It is good to give credit to the students who bring in the information or write questions. These students are then encouraged to continue looking, writing, etc.

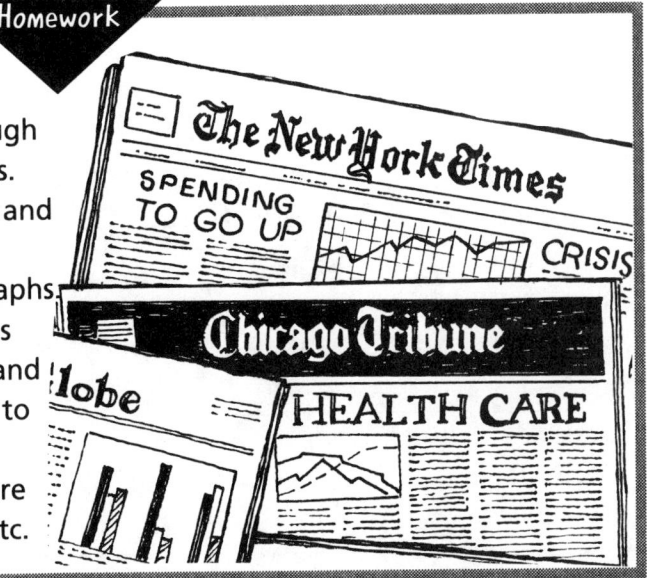

Predict and Experiment

Make two spinners with six equal spaces. On Spinner 1, write the numbers from **1** to **6**. On Spinner 2, write **1** on three spaces, **2** on two spaces, and **3** on one space.

Spinner 1   Spinner 2

If you were spinning 12 times, predict the number of times you would probably land on each number on Spinner 1.

Prediction 1. _____ 2. _____ 3. _____ 4. _____ 5. _____ 6. _____

Actual 1. _____ 2. _____ 3. _____ 4. _____ 5. _____ 6. _____

Try this twice more. Average your spins.

2nd try 1. _____ 2. _____ 3. _____ 4. _____ 5. _____ 6. _____

3rd try 1. _____ 2. _____ 3. _____ 4. _____ 5. _____ 6. _____

Average 1. _____ 2. _____ 3. _____ 4. _____ 5. _____ 6. _____

Compare your results with your class members. Average the class totals on the board with your teacher's help.

Class average 1. _____ 2. _____ 3. _____ 4. _____ 5. _____ 6. _____

Compare the class average with your predictions. Discuss why scientists need to do many experiments before they announce results.

Spinner 2—Write how many times out of 12 spins you predict landing on:

1. _____ 2. _____ 3. _____

I chose these because _____

_____

Actual Results   First try   1. _____ 2. _____ 3. _____

Second try   1. _____ 2. _____ 3. _____

Third try   1. _____ 2. _____ 3. _____

Average   1. _____ 2. _____ 3. _____

Did your results and the class results turn out the way you thought they would?

_____

_____

Name_____

What's On?

You can find out many things from charts and schedules.
Read the television schedule below and answer the questions.

Saturday Morning TV Guide

Channel	6:00 a.m.	6:30 a.m.	7:00 a.m.	7:30 a.m.	8:00 a.m.	8:30 a.m.
2 KCBS	Field Trip	Growing Up	Animal Fun	Puppy Pride	Cars and Trucks	Turtle Time
5 KTLA	Dragon Ball	Mice From Mars	Today in U.S.A.	Fantastic Fred	Giant Spiders	Harry the Hippo
7 KABC	Nosey News	Call It	Strong Man	Wonderman	Candy	Time for Fun
9 KCAL	Creature Fun	Girl Time	Adventures of Wild Warriors	Sing Me a Song		Feed the Children
11 KTTV	Jellybean Jam	News Time	Masked Ranger	Tommy's Castle	Ranger Rovers	Attack of Killer Tomatoes
28 KCET				Reading for Fun	Friendly Fun	Where Is Jimmy?

1. What time does the first show start? _____
2. What time will you need to be awake to watch "Growing Up?" _____
   What channel is it on? _____
3. What time will you be able to see "Adventures of Wild Warriors?" _____
4. What show follows "Adventures of Wild Warriors" on the same channel? _____
   How long will it last? _____
5. What channel must you watch to learn more about reading? _____
6. Why wouldn't you already be watching that channel? _____
7. At what time do you change channels to see "Turtle Time?" _____
   What channel is it on? _____

Graphs, Graphs, Graphs

There are many different ways to show information on graphs. Read each graph below and answer the questions.

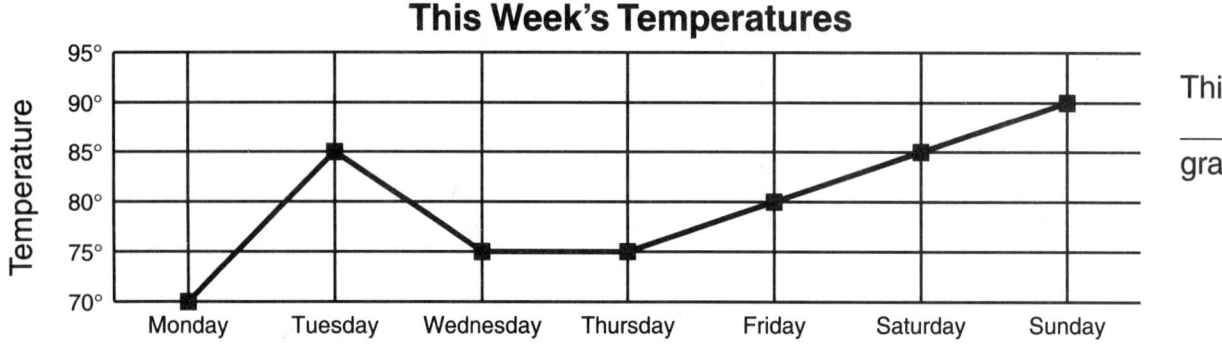

This is a _____ graph.

1. _____ was the warmest day. It was _____ degrees.
2. What two days had the same cool temperature? _____
3. Is the temperature rising or going down? _____
4. What day was the coolest day? _____
   What was the temperature? _____
5. What was the difference in temperature between Monday and Sunday? _____
6. What two days had the same warm temperature? _____

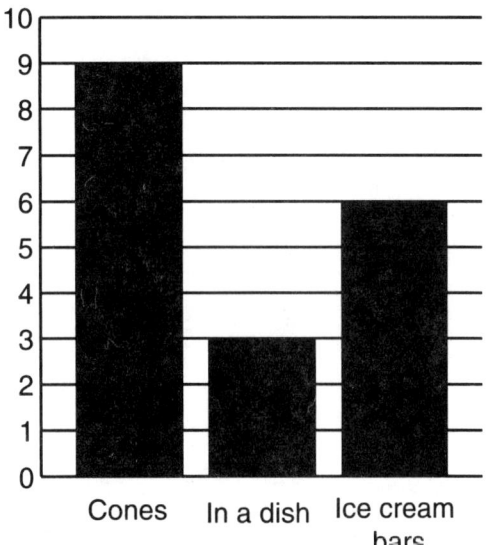

7. How many more people prefer cones to bars? _____
8. How many less people prefer a dish of ice cream than a cone? _____
9. What do people prefer the most? _____
10. How many people were polled? _____
11. How many people like cones and bars? _____
12. How many people like a dish of ice cream and bars? _____

This is a _____ graph.

Name_____

Favorite Fun

Favorite Classes

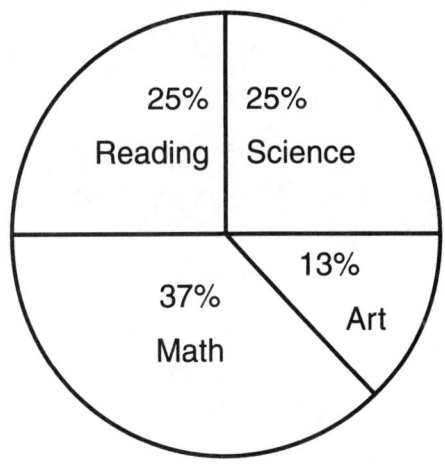

Use the graph to answer the questions below. The graph represents 32 fourth-graders' favorite classes.

1. What class is liked the most? _____

   What is the percent? _____

2. What two classes are liked the same? _____

   _____

   What is the percent? _____

3. What is the total percent of all the classes? _____

4. What class is liked the least? _____

5. How many fourth-graders liked reading the best? ____

This is a _____ graph.

Make a line graph and a bar graph using the information below. Don't forget to label each side of the graphs and then give the graphs titles.

Fourth-graders were asked what their favorite type of book to read was. The response was as follows:

    12 liked sports.

    8 liked romance.

    10 liked history.

    5 liked mystery.

    5 liked riddles.

Name________________________________

Riddle

No Bones About It!

What is the largest bone in your body?

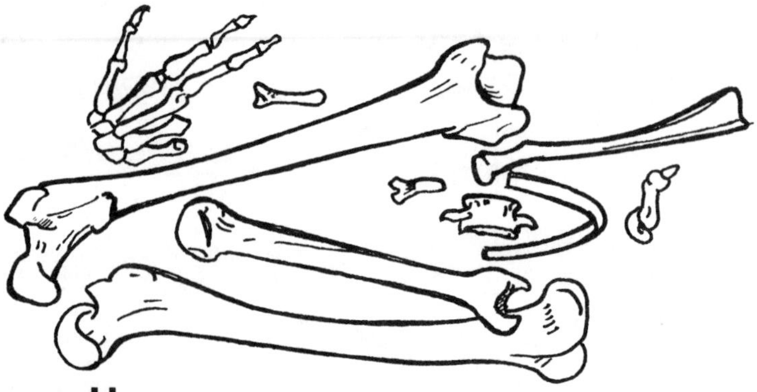

Find the points at the bottom of the page on the graph. Write the letter that represents each point above its ordered pair to find the answer to the riddle.

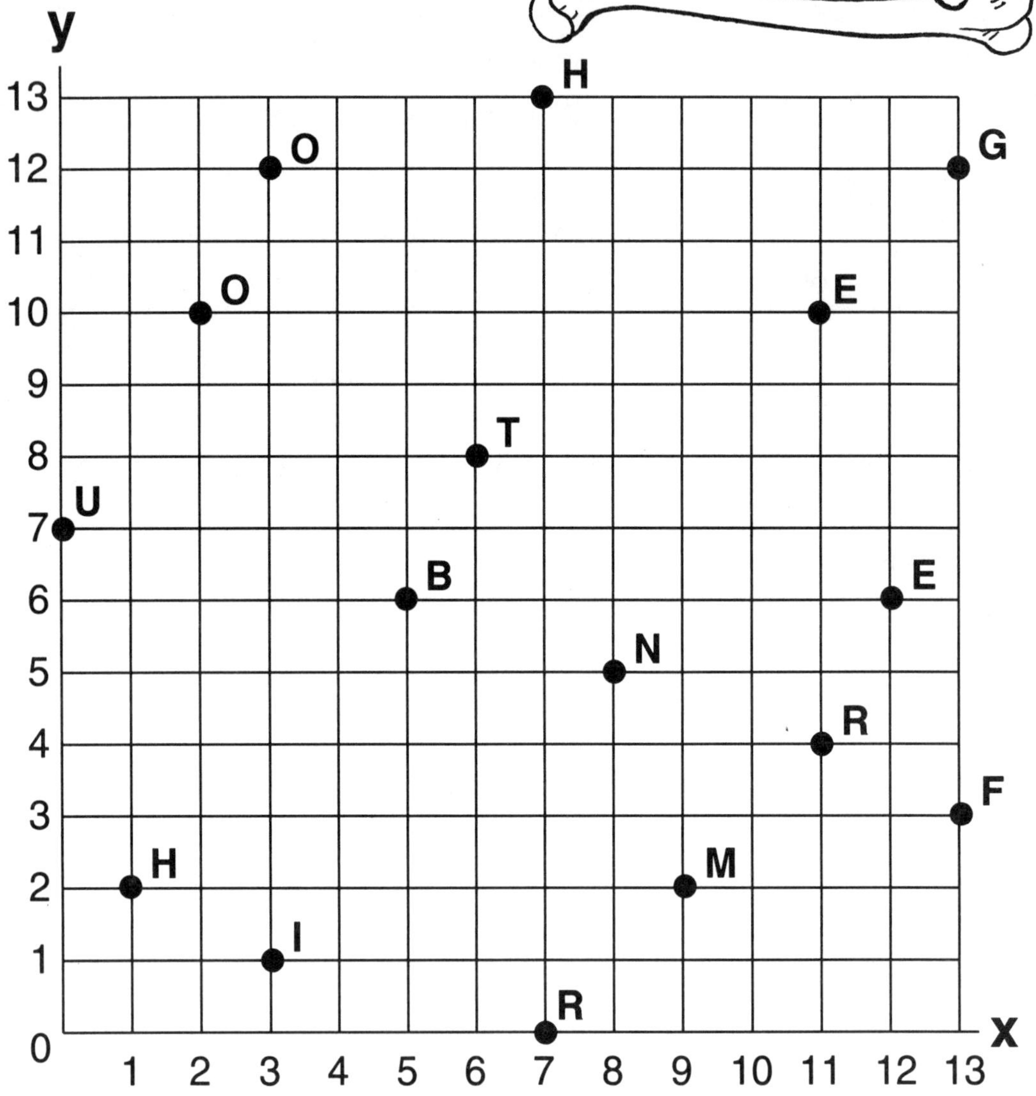

___ (6, 8) ___ (1, 2) ___ (3, 1) ___ (13, 12) ___ (7, 13) ___ (5, 6) ___ (3, 12) ___ (8, 5) ___ (12, 6)

(___ (2, 10) ___ (11, 4) ___ (13, 3) ___ (11, 10) ___ (9, 2) ___ (0, 7) ___ (7, 0))

Page 3

Millions			Thousands			Ones		
		6	5	7	0	3	5	1	6
	8	4	1	6	5	4	8	1	2
	1	4	0	3	7	2	1	6	6

(rows 1, 2, 3)

4. 5,700,902
5. 20,057,684
6. 1,906,206
7. 2,640,321
8. 9,202,000
9. 25,000,801
10. 19,050,905
11. 403,090,077
12. 17,064,096

Page 4

(crossword puzzle)

2. eight hundred twenty-one million, three hundred thirty-seven thousand, ten
3. thirteen thousand, nine hundred eighty-six
4. one million, one hundred forty-five thousand, five hundred three
5. two million, three hundred nine thousand, six hundred forty-nine
7. one hundred seventy-eight thousand, thirty-two
12. eight million, five hundred ten thousand, five hundred thirty-seven
13. eight hundred nineteen thousand, one
15. twenty-five thousand, eighty-one
17. seventy-two thousand, one hundred twenty-three

Page 5

A. <, >, <
B. >, >, <
C. >, <, >
D. <, >, >

Page 5 (E–G)

E.
1. 55,806
2. 63,744
3. 87,214
4. 92,813
5. 192,874
6. 216,606
7. 378,461
8. 471,622
9. 694,213
10. 756,802
11. 820,000
12. 971,246

F. 55,806; 971,246
G. 55,796; 55,906; 972,246

Page 6

A. 860; 50; 290; 60; 330
B. 500; 800; 300; 1,400; 5,600
C. 6,000; 8,000; 10,000; 2,000; 1,000
D. The numbers in boldfaced type should be colored.
1. **5,000**
2. **10,000**
3. 7,000
4. **3,000**
5. **1,000**
6. 3,000
7. 6,000
8. **6,000**
9. 1,000
10. 4,000
11. **8,000**
12. **7,000**
13. 2,000
14. 4,000
15. **9,000**

E. one; two; three

Page 7

A. 5, 7, 8, 20, 30, 70, 80, 200, 300, 350, 500, 750, 800
B. XI, XII, XIII; XV, XVI, XVII, XVIII; XIV; XIX
C. 22, 33, 52, 491, 150, 160, 550, 651

D.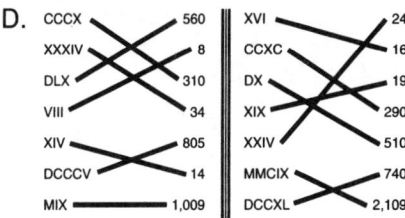

Page 11

(puzzle answers)
1,137; 61,114; 100,011; 857
b; y; y; b
10,616; 11,338; 15,996
g; r; g; r; g
r; r
4,766; 1,452; 1,474; 7,742
y; b; y
37,422; 810; 63,256

1. 987
2. 1,412

Page 12

1. 13,943 > 10,129; hexagon
2. 8,554 > 6,645; square
3. 6,343 > 5,132; triangle
4. 13,237 > 12,141; circle
5. 7,437 < 12,191; pentagon
6. 17,245 < 18,104; rhombus

Page 13

A. baseballs = $3.89, $7.78; bat = $11.64, $11.64; glove = $19.99, $19.99; **$39.41**
B. racket = $29.95, $59.90; balls = $3.25, $9.75; **$69.65**
C. skates = $42.75, $42.75; balls = $3.25, $6.50; **$49.25**
D. footballs = $12.84, $25.68; baseballs = $3.89, $11.67; balls = $3.25, $13.00; **$50.35**

Page 14

A.
1. 64
2. 36
3. 288
4. 2,901
5. 2,935
6. 326
7. 5,046
8. 1,337
9. 48,895

B.
1. 168
2. 3,729
3. 4,703
4. 241
5. 4,433
6. 655

smallest to largest = 168; 241; 655; 3,729; 4,433; 4,703

Page 15

1. 3,058
2. 2,042
3. 3,957
4. 1,101
5. 3,876
6. 1,010
7. 2,389
8. 1,111
9. 19,136
10. 31,567
11. 220
12. 3,068
13. 35,127
14. 2,769
15. 18,798

Page 16

A. $3.45 + $1.75 + $1.24 = $6.44, change $3.56
B. $4.38 + $2.04 = $6.42, change $3.58
C. $11.74 + $3.32 + $2.34 = $17.40, change $2.60
D. $9.24 + $1.02 + $2.16 = $12.42, change $2.58

ANSWER KEY

Page 17

275 + 759 = 1,034 + 9,691 = 10,725 − 1,329 = 9,396 − 4,903 = 4,493 + 759 = 5,252 + 3,308 = 8,560 − 1,329 = 7,231 − 759 = 6,472 + 4,903 = 11,375 − 9,691 = 1,684 + 759 = 2,443 + 759 = 3,202 + 4,903 = 8,105 − 3,308 = 4,797 + 1,329 = 6,126 − 4,903 = 1,223 + 9,691 = 10,914

Page 18

1. 500   2. 1,000   3. 300
4. 300   5. 800   6. 400
7. 200   8. 300   9. 600
10. 600   11. 300   12. 1,200
13. 400   14. 0   15. 400
16. 1,000   17. 1,000   18. 300
19. 700   20. 300   21. 100
22. 200   23. 600   24. 100

Page 19

1. 274; 1,692
2. 5; 363
3. $17.51; $2.49
4. 1,109; 7,347
5. $38.40; $9.09
6. 13,717; 41,689; 22,031

Page 24

Page 25

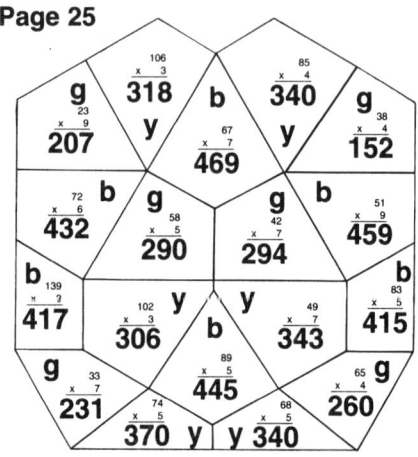

Page 26

A. 1,659; 2,278; 3,420; 464
B. 11,457; 3,838; 31,500; 55,488
C. 4,320; 2,450; 19,416; 2,400
D. 38,440; 6,072; 1,232; 1,134

Page 27

GREAT JOB!

Page 28

1. 5, 8, 7, 9, 6
2. 2, 9, 9, 8, 8
3. 9, 3, 7, 9, 4
4. 8, 2, 8, 3, 6
5. 5, 9, 5, 2, 4
6. 5, 4, 8, 3, 7
7. 2, 5, 3, 6, 5
8. 7, 2, 8, 8, 4

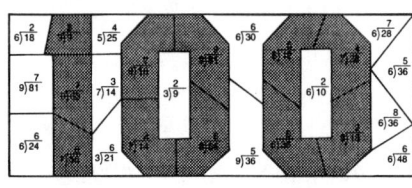

Page 29

1. 60   2. 41   3. 50
4. 80   5. 40   6. 91
7. 6 R6   8. 5 R7   9. 9 R3
10. 7 R3   11. 7 R2   12. 5 R7

60	41	50	80	30	70	91	13	14
40	4R6	5R1	40	91	6R6	8R2	9R1	3R2
70	18	24	5R1	4R2	5R1	7R4	8R2	9R3
5R5	7R3	6R5	5R2	9R1	9R3	7R2	5R7

Page 30

(crossword-style grid answer)

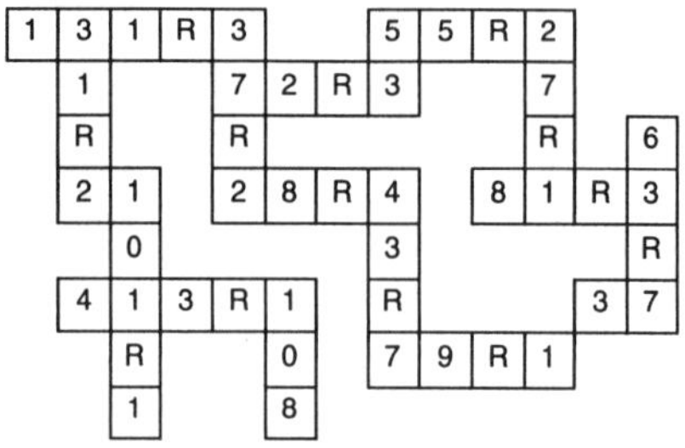

Page 31

1. 304 R2   2. 109 R2
3. 809   4. 101 R3
5. 160   6. 108 R3

Order: 101 R3, 108 R3, 109 R2, 160, 304 R2, 809

7. 0, 5, 2
8. 0, 8, 8, 0
9. 0, 3, 2, 0, 3
10. 0, 8, 1, 8, 8
11. 7, 2, 4, 2, 1, 0
12. 0, 4, 5, 5, 9, 4
13. 8, 2, 1, 6, 0
14. 0, 5, 1, 4, 3, 5, 3, 5, 0

Page 32

YOU DID IT!
Sue Ann—90, 18
James—75, 15
Tommy—95, 19
Carlos—86, 17.2
Anna—98, 19.6
Anna, James

Page 33

85 R17	30 R29	14 R27	232 R3	107 R4
23)1972	32)989	43)629	30)6963	65)6959
104 R8	31 R4	31 R12	79 R6	155 R11
15)1508	22)687	19)582	87)6933	52)8071
255 R15	14 R15	20 R18	54 R9	31 R3
31)8013	42)6231	26)541	18)819	26)809
52 R27	158 R31	36 R11	115 R8	58 R3
60)3207	39)6193	27)983	71)8203	17)989
	18 R26	44 R7	63 R9	
	81)1484	44)1943	28)1773	

Page 38

1. 0.3   2. 0.6
3. 0.9   4. 0.5
5. $8/10 = 0.8$ = eight tenths;
   $1/10 = 0.1$ = one tenth;
   $7/10 = 0.7$ = seven tenths;
   $4/10 = 0.4$ = four tenths
6. 1.4   7. 2.1
8. 3.6
9. 1.22; 3.03; 4.31

Page 39

$6\ 9/10$, six and nine tenths; $4\ 73/100$, four and seventy-three hundredths; $84/100$, eighty-four hundredths; $7\ 4/10$, seven and four tenths

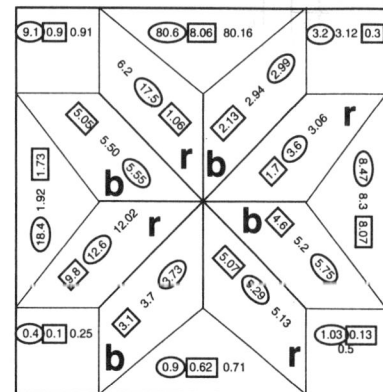

Page 40
A. 5.5, 2.8, 9.16, 0.95, 5.45, 21.8
B. 22.8, 98.3, 179.84, 71.6, 14.56, 28.30
C. 20.98, 15.3, 5.67
D. 3, 4, 0; 7, 9, 1; 4, 4, 4; 1, 3, 8; 7, 9, 3
E. 4, 3, 3; 1, 0, 3; 6, 8
F.

4.9, 6.2	1.7, 1.6	12.06, 13.2
12.4, 9.6	8.0, 8.01	2.7, 2.67
5.01, 5.35	2.66, 2.65	72.9, 79.2
2.13, 2.03	1.6, 1.7	5.00, 4.9
2.04, 2.14	4.01, 4.00	18.6, 18.65
3.1, 3.01	9.9, 10.01	2.75, 2.70

Page 41
1. 99.26
2. 33.88
3. 471.62
4. 56.42
5. 169.13
6. 84.35
7. 77.49
8. 20.95
9. 731.43
10. 86.4
11. 682.43
12. 509.22

Page 42
1. 38
2. 0.047
3. 400
4. 52,100
5. 0.062
6. 0.081
7. 55,300
8. 0.00082
9. 90
10. 0.00676
11. 33,700
12. 0.13

Page 43
44, 16 = 4; 30, 12 = 6; 40, 24 = 8; 25, 35 = 5; 22, 33 = 11; 6, 21 = 3

Page 44
GRAND PANTANAL

Page 45
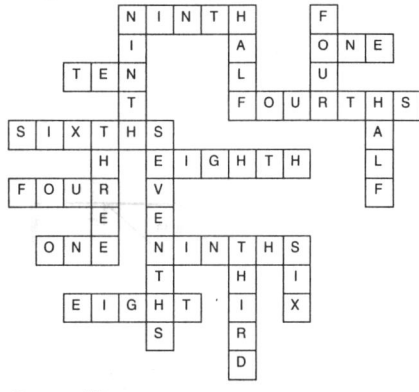

Page 46
1. 2, 2, 2
2. 2, 2, 4
3. 2, 2, 2
4. 3, 3, 6
5. 6, 6, 6
6. 4, 4, 8

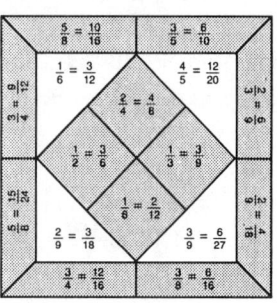

Page 47
1. $2\frac{1}{8}$
2. $\frac{5}{4}$
3. $3\frac{1}{4}$
4. $\frac{11}{5}$
5. $5\frac{2}{9}$
6. $\frac{22}{5}$
7. $3\frac{3}{4}$
8. $\frac{32}{9}$
9. $\frac{41}{9}$
10. 5
11. $\frac{41}{8}$
12. 4
13. $\frac{79}{9}$
14. $7\frac{4}{9}$
15. $6\frac{1}{6}$
16. $\frac{38}{5}$
17. $16\frac{3}{5}$
18. $5\frac{4}{7}$

Page 48
A. >, =, >
B. <, >, <
C. =, <, >
D. >, =, >

Page 49
A. $\frac{4}{5}$; $\frac{5}{8}$; $\frac{3}{4}$; $\frac{2}{8}$
B. $\frac{4}{8}$; $\frac{7}{9}$; $\frac{2}{4}$; $\frac{4}{12}$
C. $\frac{6}{12}$; $\frac{8}{8}$; $\frac{6}{9}$; $\frac{1}{3}$

Page 50

$\frac{2}{3}$	+	$\frac{14}{15}$	=	$1\frac{3}{5}$
+		+		
$\frac{1}{30}$	+	$\frac{1}{6}$	=	$\frac{1}{5}$
=		=		+
$\frac{7}{10}$		$1\frac{1}{10}$		$\frac{1}{40}$
				=
$\frac{1}{8}$	+	$\frac{1}{10}$	=	$\frac{9}{40}$
+		+		
$\frac{1}{4}$	+	$\frac{1}{5}$	=	$\frac{9}{20}$
=		=		+
$\frac{3}{8}$	+	$\frac{3}{10}$	=	$\frac{27}{40}$
				=
				$1\frac{1}{8}$

Page 51
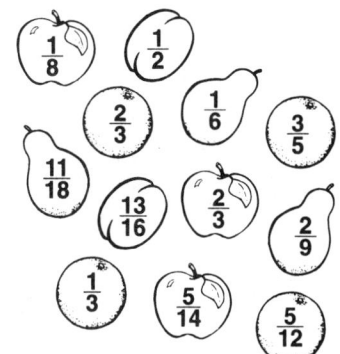

Page 55
1. triangle
2. square
3. hexagon
4. pentagon
5. rectangle
6. octagon
7. trapezoid
8. parallelogram
9. quadrilateral
10. cube
11. pyramid
12. cylinder
13. sphere
14. cone
15. rectangular prism
16. Pictures will vary.
17. face
18. vertex
19. edge
20. face
21. 3, 3
22. 4, 4
23. 6, 6

Page 56
A.

B.

C. Figures will vary.
D. 1. B
   2. AB, BD, BC
   3. CD
   4.

Page 57
1. A ∥ B
2. 1, 2, 3, 4, 8, 9, 11, 12
3. 1 and 2, 3 and 4, 5, 6, 7, 8 and 9, 10, 11 and 12
4. B ⊥ C, A ⊥ C
5. acute
6. obtuse
7. C
8. right
9.
10.
11.

Page 58
ANGEL FALLS

ANSWER KEY

Page 59
1. 4   2. 6   3. 5
4. 8   5. 8
6. 6 x 4 = 24 ft.²
7. 5 x 5 = 25 ft.²
8. 16 ft.²   9. 36 ft.²
10. 90 ft.²   11. 36 ft.²
12. 75 ft.²   13. 22 ft.²
14. 81 ft.²

Page 60
FORTY-THREE

Page 61
1. 12   2. 6   3. 3
4. 2 x 1 x 3 = 6 cu. units
5. 1 x 2 x 2 = 4 cu. units
6. 3 x 2 x 2 = 12 cu. units
7. 2 x 3 x 2 = 12 cu. units
8. 4 x 2 x 3 = 24 cu. units
9. 1 x 1 x 4 = 4 cu. units

Page 64
A. cm or mm   B. cm or mm
C. cm or mm   D. km
E. km   F. mm
G. cm   H. mm
I. g   J. kg or g
K. kg   L. g   M. mL
N. mL   O. L   P. L
Q. mL   R. L

Page 65
1. inches   2. inches
3. miles   4. yards
5. yards or feet   6. feet
7. ounces, pounds
8. ounces   9. ounces
10. pounds   11. ounces
12. tons   13. cup
14. gallon   15. gallon
16. quart   17. pint or quart
18. pint

Length:
metric—1. cm, 2. mm, 3. km;
standard—1. in., 2. ft., 3. yd., 4. mi.

Weight:
metric—1. g, 2. kg;
standard—1. oz., 2. lb., 3. T

Liquid:
metric—1. mL, 2. L;
standard—1. o., 2. pt., 3. qt., 4. gal.

Page 66
Water boils.—212°F, 100°C
Normal body temperature—98.6°F, 37°C
A very hot day—104°F, 40°C
A warm day—86°F, 30°C
A cool day—50°F, 10°C
Water freezes.—32°F, 0°C

Page 70
Answers will vary.

Page 71
1. 6:00 a.m.   2. 6:30 a.m., 2
3. 7:00 a.m.
4. "Sing Me a Song," 1 hour
5. 28
6. Doesn't come on air until 7:30 a.m.
7. 8:30 a.m., 2

Page 72
Line:
1. Sunday, 90°
2. Wednesday, Thursday
3. rising
4. Monday, 70°
5. 20°
6. Tuesday, Saturday
Bar:
7. 3   8. 6
9. cones   10. 18
11. 15   12. 9

Page 73
Circle:
1. math, 37%
2. reading, science, 25%
3. 100%   4. art   5. 8

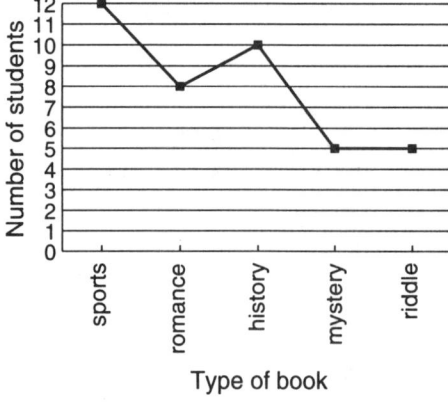

Page 74
THIGH BONE (OR FEMUR)